Time and Timelessness

Time and Timelessness examines the development of Jung's understanding of time throughout his opus, and the ways in which this concept has affected key elements of his work. In this book Yiassemides suggests that temporality plays an important role in many of Jung's central ideas, and is closely interlinked with his overall approach to the psyche and the cosmos at large.

Jung proposed a profound truth: that time is relative at large. To appreciate the whole of our experience we must reach beyond causality and temporal linearity, to develop an approach that allows for multidimensional and synchronistic experiences. Jung's understanding surpassed Freud's dichotomous approach, which restricted timelessness to the unconscious; his time theory allows us to reach beyond the everyday time-bound world into a greater realm, rich with meaning and connection.

The book looks at:

- Jung's time theory
- the death of time
- time and spatial metaphors
- the role of time in precognition, telepathy and synchronicity
- *Unus mundus* and time
- a comparison of Freud's and Jung's time theories: temporal directionality, dimensionality, and the role of timelessness.

This book is the first to explore time and timelessness in a systematic manner from a Jungian perspective, and the first to investigate how the concept of time affected the overall development of Jung's theory. It will be key reading for psychoanalytic scholars and clinicians, as well as those working in the field of phenomenological philosophy.

Angeliki Yiassemides is a developmental and analytical psychologist working in private practice in Nicosia, Cyprus.

Research in Analytical Psychology and Jungian Studies Series

Series Advisor: Andrew Samuels

Professor of Analytical Psychology, Essex University, UK.

The *Research in Analytical Psychology and Jungian Studies* series features scholarly works that are, broadly speaking, of an empirical nature. The series comprises research-focused volumes involving qualitative and quantitative research, historical/archival research, theoretical developments, heuristic research, grounded theory, narrative approaches, collaborative research, practitioner-led research and self-study. The series also includes focused works by clinical practitioners, and provides new research-informed explorations of the work of C. G. Jung that will appeal to researchers, academics and scholars alike.

Books in this series:

Time and Timelessness
Temporality in the theory of Carl Jung
Angeliki Yiassemides

Time and Timelessness

Temporality in the theory of Carl Jung

Angeliki Yiassemides

Routledge
Taylor & Francis Group
LONDON AND NEW YORK

First published 2014
by Routledge
27 Church Road, Hove, East Sussex, BN3 2FA

Simultaneously published in the USA and Canada
by Routledge
711 Third Avenue, New York, NY 10017

Routledge is an imprint of the Taylor & Francis Group, an informa business

British Library Cataloguing in Publication Data
A catalogue record for this book is available from the British Library

Library of Congress Cataloging in Publication Data
Yiassemides, Angeliki.
Time and timelessness : temporality in the theory of Carl Jung / Angeliki Yiassemides.
pages cm
Includes bibliographical references. 1. Jung, C. G. (Carl Gustav), 1875-1961. 2. Jungian psychology. 3. Time--Psychological aspects. I. Title.
BF109.J8Y53 2014
150.19'54--dc23
2013002609

ISBN: 978-0-415-81043-2 (hbk)
ISBN: 978-0-203-07099-4 (ebk)

Typeset in Garamond
by Saxon Graphics Ltd, Derby

To my boys, Petros and Ektoras,
for giving me time and for opening the gateway to timelessness.

Out of a timeless world
Shadows fall upon Time,
From a beauty older than earth,
A ladder the soul may climb.
I climb by the phantom stair
To a whiteness older than Time.

(Russell 'A. E.', *Voices of the Stones*, 1931)

Time and Timelessness, by Melita Couta, Ink and Collage on paper, 30cm x 42cm, 2012.

Contents

Preface

> Alice sighed wearily. 'I think you might do something better with the time,' she said, 'than wasting it in asking riddles that have no answers.'
> 'If you knew Time as well as I do,' said the Hatter, 'you wouldn't talk about wasting *it*. *It*'s *him*.'
> 'I don't know what you mean,' said Alice.
> 'Of course you don't!' the Hatter said, tossing his head contemptuously. 'I dare say you never even spoke to Time!'
>
> Lewis Carroll, *Alice's Adventures in Wonderland*[1]

Time is an extremely obscure concept. It is also an indispensable element of our experience: it is the stage upon which our lives unfold and end; death is identified as 'the end of time'. And although many philosophers, scientists and thinkers have attempted to define it, nobody has succeeded in pinning it down and articulating a definitive theory about its nature. Claims of 'knowing Time' can only be made by the Mad Hatter; it appears to be a riddle with no answer – at least as of yet.

We must appreciate that time is more than a concept that needs to be scientifically and philosophically explored. There is always an implicit (if not explicit) assumption regarding time behind almost every theory concerning the nature of human existence. Each field, approach or statement needs to rest against a temporal background; each uttered sentence assumes a temporal frame. It is as if time is the barometer for every human expression.

This book does not attempt to find the answer to the riddle of Time. Rather, we step outside the rabbit hole and take a trip down a different path, which may prove to be equally perplexing. I endeavour to explore the course that Carl Gustav Jung followed to address this elusive but central phenomenon of our human existence. Although he did not intentionally develop a theory of time as such, I discern his temporal assumptions and understanding, interwoven within the fabric of his opus.

Our perception of time has evolved since the beginning of history through a dialectic process between man and nature, individual and culture. For archaic humans, time was a lived experience that encompassed their entire existence. The inexorable rhythms of nature inspired the imagination of ancient civilizations that included time in their cosmogony; there was a close relationship between time and the creative energy of the world. The ancient Greeks identified eternal time with Chronos (*Χρόνος*). Chronos' divinity might be questioned but his creative force is indisputable. Chronos 'produced from within himself the calm Aither and likewise Chaos, the empty space which had no firm bottom and was filled with darkness' (Kerényi 2006: 114).

Even though it is quite common to come across mentions of 'Kronos the god of Time' this is a misconception. Chronos, the personification of time, was not identical with Kronos the Titan who ate his own children. Nonetheless, Kronos' father Uranus, otherwise known as Aion (or Aeon), was associated with eternity; Aeon happens to be a common alternative name for Time. There is one more interesting time-related association between Chronos and Kronos: Chronos (i.e. eternal time) gave birth to the hermaphroditic god Phanes, who created the parents of the Titans, Ouranos (Heaven) and Gaia (Earth). Notwithstanding this association with temporality, Chronos and Kronos refer to two different entities and are not etymologically related.

It has been presumed that this confusion is caused by the 'rather shaky foundation of an academician's pun, long post-dating the time at which the Greeks could have been called "unsophisticated" in their cosmological beliefs' (Gell 2001: 31). I suppose that this misunderstanding could also be due to the fact that the terrible qualities of Kronos happen to match our perception of time: it devours us; its passage causes our annihilation. Indeed for many ancient civilizations time was believed to be a demolishing god.[2]

Time's relentless flow meant that, beyond the mythic dimension, there was also practical need to understand and measure it. For the ancient people life revolved around the regular cycles of the sun and moon rise, the motion of the skies and the seasons. There was a time for harvest, a time for retreat, a time for birth, a time for death. A beautiful example of the marriage between the cyclicality of time and its mythic dimension can be found in the ancient Mexican civilizations where time had concurrently a numinous and a practical aspect. The Aztecs used two calendars, each corresponding to one of the temporal dimensions: the year-calendar, which followed the quantitative aspects of time, and the *Tonalamatl* calendar, which expressed its qualitative facet.[3]

With the development of civilizations time remained a central issue of discussion for philosophers and scientists alike. Some attempted to delineate its qualities and even abolish it all together. According to Archimedes (c.287–c.212 BCE) an understanding of physics presupposes the elimination of time, which is replaced by a geometrical theory of nature. This seems to be the same principle that was followed by Galileo Galilei (1564–1642) who represented time by a straight line in order to formulate 'the law of inertia'.[4] This approach was adopted by Isaac Newton (1643–1727) who used a geometrical line to describe measurable time and supported an arithmetical explanation: 'Absolute, true and mathematical time, of itself, and from its own nature, flows equably without relation to anything external' (Newton 1952: 3). According to the premise that time is absolute, the interval between two events could be measured precisely and unambiguously and this measure would be constant, provided a good chronometer was used. Newton would concretize the clock of cosmic time for centuries to come.

Albert Einstein, at the dawn of the nineteenth century, was to break the spell of absolute time and reintroduce the notion of relative time. He observed temporal and spatial indications to be relative to the motion of the observer; individual frames of reference determine the nature of time. In a sense, according to the theory of relativity, each observer is a different clock; each individual is a reality-determining location. For Einstein, the human perception of time as having a past, present and future was solely a function of consciousness, of the human mind.

Einstein's observer, that is, the individual perspective upon which the nature of time (and space) depend, had been considered an essential factor in the understanding of temporality since the beginning of philosophy. Time was interwoven with the workings of the human experience and consciousness. For example, Heraclitus (c.535–c.475 BCE) postulated the relative nature of time by arguing that the world represents the totality of events and not of things. Aristotle (384–322 BCE) linked time with human consciousness by suggesting that without awareness of change there would be no awareness of time; time is related to the movement of the soul. For Saint Augustine (354–430) time is an experience of the mind; a true measure of time is an inner measure. According to Gottfried Wilhelm Leibniz (1646–1716), one of Newton's contemporaries and adversaries, time cannot be experienced directly since absolute time is non-existent; time is an order of successions of experiences. Even Newton, who adamantly supported the existence of absolute time, noted that there is yet another type of time, 'common time' that is experienced by humans and is commonly used instead of 'true time'. Immanuel Kant (1724–1804) explained time as a construct of experience and rejected the idea that internal time has any relation to absolute reality; time is an *a priori* form of sensory perception.

The debate over absolute versus relative time is ongoing and is far from being resolved. The division is not only an issue that affects philosophical discussions. Each time-view has important implications for the perception of the world. Subscribing to absolute time implies a linear understanding. Events occur on a timeline that is characterized by temporal demarcations: past, present, future; hence, Newton's time (similar to his predecessors) was based on a geometrical line. Also, according to the absolute view, since time is based on a particular 'line' it can be determined (assuming there is an accurate time-measuring instrument). That is, if you know what happened 'before' you are in a position to form a prediction about what will happen 'after'. Consequently, this sort of time relies on causality. There is a fixed relation between events. There is a specific before and a particular after, a yesterday and a tomorrow in a way that might be perceived as development (i.e. happening 'with time'). An absolute view of time implies the existence of 'logical' meaning and sequence of temporal demarcations, which are based on a subjective, fixed and shared (at least among members of the same group) understanding of temporality.

Relative time, on the other hand, is associated with the opposite implications. It does not mean lack of time; it simply entails *non-absolute time*. Linearity is not valid and instead a non-linear model is assumed. (This does not imply a particular shape, such as a circle; non-linear does not mean cyclical, it simply means that time is not based on a line.) Hence, causality is no longer a viable consideration: when a sequence or a linear progression cannot be determined it is impossible to establish causes and effects. Also, non-linearity allows for a multidimensional approach. As an extension of this, we could conceptualize various dimensions of time that exist in combination with other concepts, such as Einstein's space-time relativity; time and space are multidimensional (and, by default, non-linear).

Psychoanalytic writers have indicated that we live concurrently in absolute and relative time. This becomes apparent within the psychoanalytic situation. Even though external reality is ruled by a logical sequence of past-present-future whereupon life is transliterated, in the psychoanalytic setting time is relative.[5] The timeline has no pertinence; the past is re-experienced as the present, and in turn the present is meshed with the past; the future as well, via wishes, anticipations, day dreams and desires assumes the role of the present. The unfolding of psychic time is subject to different rules – or even to no rules at all, as far as objective reality is concerned. These observations are largely based on Sigmund Freud's temporal understanding.

Freud considered time to be 'a subject which would merit the most exhaustive treatment' (Freud 1920: 28). Time-related postulations were present in his work since his early writings and key aspects of his theory are intimately linked to the concept. Since his early observations time had a peculiar hue in the unconscious. In *The Project for a Scientific Psychology* (1950 [1886–99]) Freud was puzzled over the fact that, even though time passes, memories do not fade and are not forgotten. He detected two sorts of time: 'time' (which he placed in quotation marks[6]) that is associated with repetition and is linked with memory, and time as a conscious experience. Temporality in the psyche was henceforth split. In *The Interpretation of Dreams* (1900) Freud established a monumental time-related premise: the processes and contents of the unconscious have no reference to time, they are timeless; the concept of time is a modality of consciousness and is part of the child's cognitive development.[7]

Freud's observation regarding the unconscious' timelessness was monumental. Yet, he admitted not having made enough use of his pioneering observations.

> Again and again I have had the impression that we have made too little theoretical use of this fact, established beyond any doubt, of the unalterability by time of the repressed. This seems to offer an approach to the most profound discoveries. Nor, unfortunately, have I myself made any progress here.
>
> (Freud 1933: 74)

Even though the topic of time attracted his interest, Freud failed to present a coherent theory. As André Green remarked, Freud's time-related theories were developed 'in a fragmentary and unsystematic way', thus even though he offered significant insights, 'he left us with a mosaic of temporal mechanisms without conceptual unification' (Green 2009: 1). Perhaps Freud's hesitation (or discomfort) regarding a formal presentation of his 'time theory' is the explanation for the fact that he recorded his observations mostly in footnotes, as well as in peripheral comments.[8]

Regardless of the lack of a coherent time theory, Freud's 'double hypothesis'[9] dominated his work. Timelessness governs the unconscious whereas consciousness functions in, and is aware of, time. Carl Jung was in the historically advantageous position (being part of the next generation of psychoanalysts) of being able to rely on Freud's fertile thoughts. However, he was not merely affected by Freud's understanding of time; his theory was a radical departure. Even though time was introduced by both Freud and Jung in connection to the unconscious, the wider context of their respective approaches differs dramatically. Jung incorporated a temporal understanding that allowed for the expansion of Freud's observations. Their individual time theories rest upon each thinker's understanding of the psyche.

By adopting a different temporal stand, Jung allowed for the elaboration of Freud's overall approach to temporality as well as of particular time-related concepts, such as the experience of the oceanic feeling, as well as the relation between conscious attention and time. Whereas for Freud the past was of paramount significance, Jung emphasized the role of the future. Meaning is detached from the past; it encompasses the totality of our existence, which inexorably includes the future. Also, whereas Freud had connected time primarily with his endeavours to understand the role of the unconscious in the mental apparatus, Jung's first deliberations on time are found in his musings regarding the connection of the soul with nature. The respective stance Freud and Jung had on time determined the unfolding of important implications, such as their views on causality, as well as the directionality and dimensionality of the psyche. Thus, Freud's 'time theory' will be used sparingly throughout this book as a comparison against Jung's position. It is my hope that this contrast will help the reader to better appreciate Jung's 'time theory'.

Jung did not make any unique discoveries concerning the nature of time; many others had observed – in cosmological accounts, mythology, poetry, literature – what he recorded. What is exceptional is the way he related time in the context of the nature and functions of the psyche into a reasonably coherent theory. Through this book, I hope that the reader recognizes that although Jung did not intentionally develop a theory of time as such, it is possible to discern a unique understanding of time, interwoven within the fabric of his opus.

Time is interlinked with Jung's overall approach to the psyche; it is an integral element of his understanding regarding the 'order of things' in the psyche and the cosmos at large. Therefore, we need to examine the concept not simply based on the facts that he provided, but rather we must investigate in some detail the development of its function in his work; the ways in which time entered his theory and what role it played in his conceptualizations. And we will be richly rewarded for doing so!

Jung's 'time theory' started to form in 1916 with the writing of *Septem Sermones Ad Mortuos*[10] (The Seven Sermons to the Dead), a period so important that he himself acknowledged that it was the precursor of all of his later accomplishments. In his work completed before the poem, time was viewed from a symbolic perspective and was represented by the libido's oppositional quality; as the flow of life and its opposing force death. It is as though Jung's understanding of time matures *through* this poetic piece. In *Septem Sermones* time and timelessness are interwoven in a wider context; in the work following this Gnostic poem there is a more established treatment of time.

In this 'second phase' we can discern the temporal bidirectionality of Jung's 'time theory'. The past anchors the psyche, which gazes hopefully into the future. Jung urged his readers to reinvent the past, to 'dream the myth onwards'. He adopted teleology as the methodology of choice since it allows for the inclusion of *a possible future dimension* where the psyche's potentialities could unfold. Jung's investment in the future dimension becomes clear in his accounts of goal-orientation and foreknowledge. This aspect of his theory directed it away from causality, and into the realm of the acausal.

Jung's re-examination of causality led him to conceptualize time as *relative*. He suggested that its behaviour in the psyche manifests via different phenomena, such as telepathy, dreams, foreknowledge, and the acausal connecting principle of synchronicity, which can be considered as the culmination of his temporal understanding. Synchronicity is one of the most widely known terms of Jungian psychology, and yet there has been limited systematic treatment of it. Moreover, despite the fact that *Chronos* is embedded in its name, there has been barely any mention of the element of time in its analysis. Nonetheless, the temporal parameter of the theory is of pivotal importance and it remains central throughout Jung's formulations. Also, time's role in synchronicity can illuminate his understanding of the transcendent psychoid nature of the psyche, which stretches its feelers into the material realm.

In order to investigate the physical dimensions and manifestations of the psyche, Jung sought help from the field of physics. Time's behaviour was a departure point as well as a point of conformity between Jung and his contemporary physicist Wolfgang Pauli – one of the pioneers of quantum mechanics – who was largely responsible for the development of synchronicity and of the *quaternio* around which the principle developed. Time's relativity, as explored through their communication, is interwoven into Jung's understanding that reaches beyond causality and the dichotomy of 'inner' and 'outer' realms.

For Jung, the inconceivable-in-its-wholeness underlying reality is undifferentiated by fixed categories. His theory tried to encompass reality in its entirety; the physical as well as the psychic realm. These two seemingly opposing domains have meaning as their point of comparison, which can be observed via synchronistic occurrences. However, even though meaning seems to be an essential condition of the principle, additional dimensions are equally important such as the role of a conscious observer and the element of time: the observer during a synchronistic occurrence is 'at the right place, at the right time' and thus is able to detect the latent meaning, inherent in the psychoid reality. We thus investigate these parameters in light of Jung's 'time theory' so as to better understand his overall approach to the psyche.

An aspect of Jung's theory that transpires throughout his work and merits attention is the interconnection of time and space. Beyond the utilization of temporal metaphors Jung, similar to most theorists and thinkers, heavily relied on spatial language in his explorations; he 'spatialized' time in the psyche. Through the use of temporal *and* spatial metaphors his theory remains connected with the relative temporality of the psyche. The space-time duality is contained by and united through meaning. This idea finds its expression in the principle of synchronicity, which transcends spatial and temporal demarcations and points the way to a unified reality.

Jung believed that there is an overarching energy that 'shapes all things'. He implicitly regarded time as a unified temporal field, which manifests in different ways depending on particular circumstances. Contrary to Freud, the emergence of consciousness did not compromise the psyche's temporality. For Jung, *time is relative at large.* We contain within us the blueprint of the whole and can thus access absolute reality by employing faculties beyond intellectual thinking and causal explanations. His approach offers a possible solution to humanity's estrangement from the primordial oneness.

Even though time is an integral element of his theory, Jung did not include it in his reflections on a unified reality. Thus, in order to better conceptualize his hypothesis, we need to amplify the temporal dimension of time-related symbols, such as the World Clock and the Uroboros (the tail-eating serpent). These images can help us appreciate his theory since they emphasize the idea that in a unified reality the past, present and future coexist and can be observed simultaneously. Through these amplifications we will grasp an essential element of Jung's 'time theory': beyond its temporal bidirectionality, it relies on non-linearity and multidimensionality.

During his initial formulations Jung made an ambiguous statement concerning temporality in the psyche: linear time, which is one and the same as change, can be defeated *with change.* To understand this paradox and its temporal implications we must steer away from eschatology and descent to the realm of the archaic, where time has a different hue. There, we can access 'mythical consciousness' and reconnect with a 'primitive' sense of time. By allowing mythical time to enter consciousness meaning can claim the central role it was destined for in Jung's theory.

Despite their seemingly mythic dimension, Jung's ideas were based on the ordinary, that is, on the everyday life experience. Arguably, it was precisely this aspect of our humanity that he deemed 'large': the prophetic dream that is altogether factual; the accidental occurrence which is ordinary in its extraordinariness; the inexplicably regular synchronistic event. Jung would amplify his observations through myth, poetry, even positivistic science, as if to grasp their precise elusiveness. And it is through this lens that I explore time's vital role in his theory – concealed from plain view even though strikingly present. It is thus my hope that through this book even though readers might not get to know time – at least not intentionally – they will, nonetheless, appreciate the manner in which Jung 'spoke to Time'.

Notes

1 Carroll 1960: 71.
2 For example in ancient Persia the time-god Zurvan was the cause of decay and death. Shiva, one of the primary deities of Hinduism, is knows as Great Time or all-devouring Time.
3 Zavala 1982: 59.
4 Galilei's Law of Inertia: 'A material point, provided no external force acts upon it, that is, provided it is left to itself, moves along a straight line and covers in equal time equal distance' (Neumann 1976: 125).
5 See for example Cohn 1957; Loewald 1962; Pollock 1971; Namnum 1972; Kafka 1972; Arlow 1986.
6 'What is it, then, than happens to memories capable of affect till they are tamed? It cannot be supposed that "time", repetition, weakens their capacity for affect, since ordinarily that fact [repetition] actually contributes to strengthening an association. Something must no doubt happen in [the course of] "time", during the repetitions, which brings about this subjugation [of the memories]' (Freud 1950: 381).
7 The same idea was reiterated in *The Unconscious*: while dealing with the 'special characteristics of the system Unconscious' he explained that 'The processes of the system Ucs. are timeless, i.e. they are not ordered temporally, are not altered by the passage of time; they have no reference to time at all. References to time is bound up, once again, with the work of the system Cs.' (Freud 1915: 187).
8 It is in a footnote that he most emphatically wrote of the immortality and timelessness of the contents of the unconscious: '[The unconscious wishes] share this character of indestructibility with all other mental acts which are truly unconscious, i.e. which belong to the system Ucs. only. These are paths which have been laid down once and for all, which never fall into disuse and which, whenever an unconscious excitation re-cathects them, are always ready to conduct the excitatory process to discharge' (Freud 1900: 553n). In *On Narcissism: An Introduction* Freud restated – in a footnote – that 'the time-factor ... has no application to unconscious processes' (Freud 1914: 96). See also Freud 1926: 142n.
9 'Freud's invention, namely of the double hypothesis of the unconscious and of its characteristic of having no knowledge of time, was unprecedented' (Green 2002: 162).
10 Henceforth to be referred to as *Septem Sermones.*

Acknowledgements

Time and Timelessness is the product of a journey that started many years ago in New York City and was completed in Cyprus, via England. There are many friends and colleagues who have helped and inspired me along the way. Some have been with me all the way: this book would not exist without the support (practical and psychological) of my husband Markos, and my parents Ritsa and Petros. I am eternally grateful for your help and understanding. Michelle Galanter, thank you for always being 'there', and for reminding me to flow.

I am indebted to Jim Sexton, Head of the Survey and Alignment Group and Supervisor of the Accelerator Operators at Cornell University's Laboratory for Elementary-Particle Physics (LEPP) for explaining and discussing with me the intricate theories of relativity and quantum mechanics and their temporal dimensions. Thank you, Jim, for insisting that I colour outside the lines. My heartfelt gratitude to the talented Melita Couta for helping me capture in an image the symbol of 'Time and Timelessness' as I envisioned it. I want to express my appreciation to Professor Renos Papadopoulos, for his invaluable comments on various topics included in this book. I would also like to thank Andrew Samuels and Paul Bishop for encouraging me to publish my work and for their feedback.

Grateful acknowledgement is made to these publishers for permission to reprint from the following previously published material:

The Journal of Analytical Psychology for permission to reproduce the following.

Bishop, P. (2008) 'The timeliness and timelessness of the "archaic": analytical psychology, "primordial" thought, synchronicity', *Journal of Analytical Psychology,* 53: 501–23.

Yiassemides, A. (2011) 'Chronos in synchronicity: manifestations of the psychoid reality', *Journal of Analytical Psychology,* 56: 451–70.

The University of California Press for permission to reproduce Rainer Maria Rilke, *Sonnets to Orpheus*, translated by C. F. MacIntyre; © 2001 C. F. MacIntyre.

Taylor & Francis for permission to quote from Jung, C. G., *Collected Works* Vol. 8, *The Structure and Dynamics of the Psyche,* Second Edition, G. Adler and R. F. C. Hull (eds and trans) 1969, Routledge & Kegan Paul.

Disclaimer

Chapter 1

Jung and time, the beginning

Asserting the largest order[1]

He took the golden Compass, prepar'd
In God's eternal store, to circumscribe
This Universe and all created things

John Milton, *Paradise Lost*[2]

Jung's core understanding of time's role in the psyche was present since the onset of his theory. However, it appears that his 'time theory' was shaped in 1916 by the writing of *Septem Sermones Ad Mortuos* (The Seven Sermons to the Dead). I therefore divide his writings into two phases: those completed before and those completed after *Septem Sermones*. In what can be characterized as the initial phase (his work completed before the writing of the poem), time was mentioned primarily at a symbolic level. As a contrast, in the work following *Septem Sermones* Jung offered important deliberations on the topic. However, even though the latter phase is characterized by significant theoretical elaborations, it contains themes and ideas that Jung had touched upon during the first period. At the early stage of his career he limited his writing to rudimentary statements on the soul's temporality, on which he did not elaborate. As his theory matured, he returned on several occasions to his initial remarks regarding the role of time in the psyche in order to offer significant insights. There is an *enlargement* of what was always there, rather than a transformation of his original position.

Like Freud, for Jung the timelessness of the unconscious was a concept of importance. A case in point is one of his first explicit mentions of time. In *Zofingia lectures* Jung suggested that the principle of life extends beyond the control of consciousness; there exists something transcendental, to which he 'boldly assign[ed] … the name of "soul"' (Jung 1897: par. 96). Since space and time are categories used for our conscious understanding and sense perception there is no use for them beyond the material realm: 'Only forces in material form move within the boundaries of space and time' (ibid.). It follows that, since the soul is not restricted by material forms, '[it] *is an intelligence independent of space and time*' (ibid.: par. 98) Thus, according to Jung's hypotheses, the soul is immortal.

Jung argued for a 'vitalist' position against a 'mechanistic' one, where he postulated the existence of a 'nonphysiological "intellectual being" or "life force" ... This life-principle, i.e. the soul ... "extends far beyond consciousness"' (Von Franz 1983: xviii). According to von Franz, this is Jung's first indirect mention of an unconscious psyche. I would indicate that this first reference to the unconscious is intimately connected with temporality.

Time before *Septem Sermones Ad Mortuos*: symbolic elaborations

During his initial formulations Jung was referring to the way time is consciously conceived: as the flow of life and its opposing forces that cause change, and ultimately death. Time was explored as a symbol of the earthly, of the mortal human conflictual energies that oppose the eternal God(s) and the unchanging essence of being. He was in accord with ancient cosmological accounts: time is an awesome force capable of claiming the role of the creator.

In the initial phase of his temporal formulations, Jung explored several symbolizations. He elaborated on various mythological and religious images that are either connected with time within a particular context, or that Jung himself linked to time. An example of the former category is Aion, a god inexorably linked to time – 'In the Mithraic religion we meet with a strange god, Aion, also called Chronos' (Jung 1911–12/1952: par. 321). He also related time with various symbols of the libido, the term he used for psychic energy.[3] Time and the libido are intimately connected: '[Aion] is a time-symbol, and is composed entirely of libido-images' (ibid.: par. 425). Above and beyond various symbolizations, he amalgamated all time symbols with the libido.[4] The factor responsible for the proposed overlap is their *opposition*: both time symbols and libido symbols contain forces that oppose each other.

> Time is thus defined by the rising and setting sun, by the death and renewal of libido, the dawning and extinction of consciousness ... So time, this empty and purely formal concept, is expressed in the mysteries through transformations of the creative force, libido, just as time in physics is identical with the flow of the energic (sic) process.
>
> (ibid.: par. 425)

The above quote is key for understanding the nature of the libido-time connection (and will be revisited in Chapter 5). It is not simply that opposition is inherent in both and they could thus be conceptualized as comparable, or even equivalent energies or concepts. In that case opposition would have been the force for which Jung attempted to find a symbolic expression via time and the libido. Rather, this quotation indicates that time is represented *by* the libido. In an attempt to grasp the elusive nature of time Jung employed various images that contain oppositional qualities; since libido is the quintessential oppositional force it was deemed to be the most appropriate symbolization for time.

Since in his earliest accounts time was treated primarily from a symbolic point of view Jung did not discuss its impact on the reality of the psyche. However, the seeds of its vital role had been planted. In the years that followed Jung, with the assistance of a Gnostic text, was to move beyond the strictly symbolic and into a theoretically elaborate understanding of temporality in the psyche.

Septem Sermones Ad Mortuos and time

When Jung separated from Freud, and during a period of his life filled with visions and fantasies, a poem came to him that he considered to have been authored by Basilides, a second-century Gnostic from Alexandria.

> In 1916 I felt the urge to give shape to something. I was compelled from within, as it were, to formulate and express what might have been said by Philemon. This was how the *Septem Sermones ad Mortuos* with its peculiar language came into being.[5]
>
> (Jung 1961: 189–90)

Septem Sermones was the only piece of *The Red Book* published during Jung's lifetime.[6] With the publication of *The Red Book* (Jung 2009) it was revealed that the poem (with minor variations from the originally published version) was found in its closing pages. When *Septem Sermones* was published privately in 1916 and in 1925, under the title 'The Seven Sermons to the Dead written by Basilides in Alexandria, the City where the East toucheth the West' Jung did not appear as the author. Evidently, there was controversy regarding his view of the publication of the poem, which stems not from its contents but rather from the style of writing (Papadopoulos 1980).[7] The poem was not incorporated in Jung's *Collected Works* and was not on public sale. In 1961, during his lifetime, it was included in the appendix of his autobiography *Memories, Dreams, Reflections*.

It is indicative that Jung's relation with this piece was quite contradictory. His initial reference to the poem was in a letter to his friend Alphonse Maeder, to whom he sent the poem as a 'token of joy' (Jung 1973: 33–4). Jung was excited about its composition and considered it to have 'far reaching associations' (ibid.). According to this account, he chose the name of Basilides of Alexandria since he felt he did not deserve credit for it. He was, nonetheless, pleased with the coming of these words into his life: 'It fell quite unexpectedly into my lap like a ripe fruit at a time of great stress and has kindled a light of home and comfort for me in my bad hours' (ibid.).

Years later, Jung appeared to be defensive when it came to the poem, mostly to fend off accusations of being a Gnostic. In his reply to Martin Buber's charges, Jung referred to *Septem Sermones* as:

> [A] sin of my youth, committed nearly forty years ago, which consists in my once having perpetrated a poem. In this poem I express a number of psychological *aperçus* in 'Gnostic' style, because I was then studying the Gnostics with enthusiasm. My enthusiasm arose from the discovery that they were apparently the first thinkers to concern themselves (after their fashion) with the contents of the collective unconscious. I had the poem printed under the pseudonym and gave a few copies to friends, little dreaming that it would one day bear witness against me as a heretic.
>
> (Jung 1952b: par. 1501)

Arguably, Jung used Basilides' voice in order to employ the Gnostic vocabulary which allowed him to talk about subject matters that were of grave importance to him – the unconscious, collective and personal, and its contents – but for which he had not yet formed a clear theory or language.[8] At the time of *Septem Sermones'* writing Jung had yet to compose his theory regarding the nature of the psyche. It is perhaps for this reason that, besides using a pseudonym, Jung wrote a poem. The poetic form allowed him to explore without restraints his novel observations and insights concerning the psyche.[9] Jung himself characterized *Septem Sermones* as 'a poetic paraphrase of the psychology of the unconscious' (Jung 1976: 571). It has even been suggested that this poem 'was the only work of Jung's that was written "entirely" by his unconscious' (Papadopoulos 1980: 303).

To understand such an esoteric piece of literature produced by a man who was later accused of psychologizing religious teachings we should approach it as an example of his *psychological* understanding. Jung greatly admired the Gnostics for being able to identify the workings of the psyche and project them into their cosmology.

> The self [as opposed to the ego] was of course always at the centre [of the psyche], and always acted as the hidden director. Gnosticism long ago projected this state of affairs into the heavens, in the form of a metaphysical drama: ego-consciousness appearing as the vain demiurge, who fancies himself the sole creator of the world, and the self as the highest, unknowable God, whose emanation the demiurge is.
>
> (Jung 1949: par. 1419)

Septem Sermones must be treated not as a metaphysical poem, but rather as a significant component of Jung's initial attempts to explore his newly formulated ideas.

Unquestionably, this was a very important period in his life, since from these experiences stemmed the creative work of the subsequent years. Jung acknowledged that all of his later accomplishments were contained in the fantasies and dreams experienced during that time (Jung 1961). Even though its terminology was hardly used afterwards, numerous emerging theories of his later psychology can be detected in this manuscript. This also seems to be one of the earliest mentions of the process of individuation, found as *Principium Individuationis* in the poem.[10] Moreover, *Septem Sermones* was the initial ground for the development of Jung's time theory.

In view of what has been outlined above, as well as what is to follow, I would suggest that by writing this piece using a poetic language Jung had the freedom to timelessly relate to the concept of time. Poetry allowed him to elaborate imagistically, rather than theoretically, his observations regarding the nature of time. To clarify: I am not implying that *Septem Sermones* is a poem about time. Rather, it deals with concepts that are within time and are defined by it, as well as by its absence. In the spheres of the universe described by Basilides, Jung formulated an initial approach to his assumptions regarding time in the psyche. *Septem Sermones* is a key text in his development. It is the first instance that time as well as timelessness are interwoven in a wider context. In this manuscript the temporal parameters of both the cosmic and the individual levels are interrelated.

In *Septem Sermones* the timeless unconscious, as previously defined by Freud, is taken to a new level. Timelessness is not a quality belonging solely to the personal unconscious, but rather it extends beyond the boundaries of the individual, to a cosmological context. In the poem time is conceptualized in terms of the entire universe. In it we encounter collective as well as individual structures that are within and beyond time. With these parameters in mind, let us examine temporality in *Septem Sermones*.

In *Septem Sermones Ad Mortuos* the souls of the dead Christians who came back from Jerusalem are asking the poet for what they did not find. The

Gnostic Basilides lets them in and starts preaching the reality of the universe. In the first and most elaborate sermon, Jung as Basilides[11] sets the scene for the universe he is about to reveal to the seekers by making an important distinction between the realms of Pleroma and Creatura. Upon these two realms rest the subsequent sermons.

Pleroma as well as Creatura are central words in Gnostic cosmology. Pleroma[12] describes the totality of the divine; it is a heavenly state that encompasses what cannot be grasped spiritually by humans. The Gnostics believed that our world is occupied by spiritual beings that self-emanated from the Pleroma.[13] According to Basilides, Pleroma has no qualities, it is everything and nothing, and yet it encompasses all qualities. He talked about the Pleroma in order to get rid of 'the delusion that somewhere within or without there is something absolutely firm and definite' (Jung, as cited in Hoeller 1982: 45[14]). Hence, everything in it is relative, even death and creation. Pleroma is infinite, as well as eternal.

> In [the Pleroma] thinking and being ceases, because the eternal is without qualities ... In the Pleroma there is nothing and everything. We ourselves, however, are the Pleroma, so it is that the Pleroma is present within us.
>
> Even in the smallest point the Pleroma is present without any bounds, eternally and completely ... The Pleroma is the nothingness which is everywhere complete and without end.
>
> (Jung, cited in Hoeller 1982: 45)

The Pleroma's eternal and infinite quality is precisely the reason why it has no other qualities. Within its infinity co-exist everything and nothing.

The nature of the Pleroma becomes clearer when examined against the defining feature of Creatura, namely distinctiveness. Creatura[15] is encompassed within the realm of the Pleroma but, paradoxically, it is not in the Pleroma, but in itself. As opposed to the Pleroma, it is fixed and certain; it has qualities and is subject to change. Whereas in the Pleroma there are no distinctions, in Creatura effects are brought about *because of change*; that alone is fixed and certain which is subject to change. Effects are brought about, there is action, there is creation. Time enters the universal strata via the realm of Creatura. There can be no change without the passage of time since change presupposes a temporal reality. It follows that temporality is what defines the difference between the two realms.

Via *Septem Sermones* Basilides offered an explanation of the temporal duality of the human nature: humans are 'infinitely' and 'essentially' removed from the Pleroma; we are in our essence part of Creatura and since we belong to it we share its quality of distinctiveness. Our fundamental nature is confined within time and space. Nonetheless, since Pleroma encompasses everything, humans too are part of it and posses its qualities. Even though humans are essentially time-and-space-bound, we simultaneously belong to the endless, boundless and eternal. We are part of the great eternal whole, as well as of a distinct and separate realm, which is defined by time.

Distinctiveness, Basilides emphasized, is an important aspect of the human nature. Whereas Pleroma encompasses everything, the quintessence of humanity is differentiation and distinctiveness: 'For this reason the created also causes further differentiation. That is why man himself is a divider, inasmuch as his essence is also differentiation' (Jung, as cited in Hoeller 1982: 46). By virtue of their nature, humans create differentiation and division in the Pleroma, where in fact there is none. Also, humans have the need to speak about differentiation so as to understand it. They must mirror their distinctive qualities in the universe and engage in discrimination by differentiating the non-existing properties of the Pleroma. If they fail to do so they move beyond their humanity and into the realm of infinity.

Basilides warned that indistinctiveness equals the death of human nature, and offered *Principium Individuationis* as a principle that could save man from annihilation resulting from lack of discrimination: 'We must be able to distinguish the qualities of the Pleroma. Its qualities are the PAIRS OF OPPOSITES' (ibid.: 27; original uppercase). These opposites are traits of the Pleroma that are also in us, since we are part of it; they are an indispensable part of human nature.

He offered some examples of pairs such as Fullness and Emptiness, Living and Dead, Time and Space and emphasized that, even though they both exist in the Pleroma, they are balanced and cancel each other out. Contrary to this Pleromatic equilibrium humans are victims of these opposites. In order to be set free we must learn to distinguish ourselves from these qualities. However, as mentioned above, there is one important peculiarity. These opposites are non-existent in the Pleroma. Our Creatura-bound existence, *our being*, is distinctive in its essence. Basilides instructed that we must not 'strive after differentiation and discrimination as [we] know these, but strive after [our] true nature' (ibid.: 48).

To summarize this paradoxical position: humans must differentiate themselves from pairs of opposites with the intention of remaining true to their distinctive nature, thus avoiding the dissolution and nothingness of the Pleroma. If we are to actualize human nature – which is bound by Creatura but belongs to the Pleroma – we must not think but rather strive after our own being. A return to the true human nature is in order, which will free humanity from the annihilating qualities of the Pleroma. Since differentiation and opposition are fundamental characteristics of our humanity, we are bound to observe the same qualities in the outer world. We create differentiation in the universe at large by projecting our inner reality. It is for this reason that the individual and the cosmic share the same fundamental structure.

Before turning to some possible interpretations of the above ideas let us examine how time enters the struggle of our psyche's return to its 'true nature'. As mentioned above, according to the poem, Time and Space are among the thought-generated pairs of opposites. Upon first glance, it seems peculiar that time is proposed to exist in opposition to space much like hot versus cold. How is time in opposition to space?

There are two possible explanations. The first has to do with the specific time in history the poem was composed. *Septem Sermones* came into being in 1916, the same year that marked the completion of the Theory of General Relativity. According to Jung's own accounts, he was greatly interested in Albert Einstein's new theorem, and often conversed with him while the great physicist was in the process of formulating his groundbreaking theory.[16] As is well known, with it Einstein introduced time as the fourth dimension by which reality could be defined. Time entered the equation alongside space. Perhaps this contributed to the inclusion of time and space among the selected pairs.

The second consideration, and perhaps the most important one for the understanding of time and space as a pair of opposites, is the Gnostic conceptualization of such pairs:

> Binaries, or Syzygies as they are more often called, are a time-honored feature of Gnosticism ... One of the chief hallmarks of these emanations is that they are represented in pairs, one male and other female, and that their names usually describe contrasting or complementary qualities of being. [In *Septem Sermones*] [t]hese opposites are ... existential realities within the sphere of human life.
>
> (Hoeller 1982: 74)

In the Pleroma these complementary qualities are non-existent, whereas in the human experience of the world they are real. Time and space are two of the primary dimensions used by humans for differentiation and, consequently, definition of their existence. In his later writings Jung would propose that time and space are two of the intellectually necessary principles – together with causality and synchronicity by which our human experience is defined. In the poem itself, time and space are not contradictory but rather they are balanced dimensions/qualities of the same concept. Pairs of opposites in the Pleroma cancel each other out. Syzygies are in perfect balance, whereas in human nature they are in action and thus out of balance.[17] Basilides explained:

> If we know how to know ourselves as being apart from the pairs of opposites, then we have attained salvation ... When you strive after differentiation or sameness or after other qualities, you strive after thoughts which flow to you from the Pleroma, namely thoughts about the non-existent qualities of the Pleroma ... Not your thinking but your being is differentiation.
>
> (Jung, as cited in Hoeller 1982: 47–8)

We must not strive to master either time or space (or any other twofold set, for that matter) by our thoughts, for doing so would result in estrangement from our true nature. Rather, we must distinguish our being from time, as well as from space, in order to avoid getting lost in the Pleromatic state, which is spaceless and timeless and encompasses all of time and space.

Symbolic precursors of analytical psychology

The psychological implications of *Septem Sermones* can be interpreted on the personal as well as the cosmic/collective level. Let us examine at this point the above concepts in light of Jung's understanding of the psyche as it was formulated following *Septem Sermones*.

The Pleroma can be seen as the original fullness, the entirety of the psychic reality. It has been suggested that it corresponds to the realm of the self. The Pleromatic realm, like the self, encompasses everything; it is the great whole where nothing is distinctive and everything exists. On the other hand, Creatura corresponds to the realm of the ego. Similar to the ego, it differentiates, distinguishes, is changeable and is bound by action. In *Septem Sermones* Jung as Basilides formulated for the first time the ego's strive in relation to the self.[18] This becomes more relevant in light of Jung's later formulations: we must not dissolve into the self; we must distinguish from its indistinctiveness otherwise we are given over to dissolution into nothingness. '[Man is] a vessel filled with divine conflict' (Jung 1952a: par. 659). Our essence is distinctiveness, and towards it we must strive.

Our psyche contains all of the potentialities and powers of the universe (which we access via archetypal images) including the image of God. By recognizing this powerful capacity and incorporating it with consciousness we can become integrated with this eternal process. As his theory developed Jung provided a detailed account regarding the process via which the differentiated ego strives to return to the original wholeness of psychic reality. In the poem we can detect the first mention of the conditions under which the ego (i.e. Creatura) must attempt such a return. The ego must stay true to its nature and always discriminate, otherwise there is great harm. Basilides warned that if the ego submerges into the Pleromatic state, 'we extend ourselves beyond the created world, and we fall into the undifferentiated state which is another quality of the Pleroma' (Jung, as cited in Hoeller 1982: 46). In order for the ego to remain alive and not perish within the realm of the self its qualities must be differentiated. Humans must accept their differential, space-and-time-bound nature, but at the same time realize their participation in the non-differentiated and eternal reality of the universe.

In the first Sermon Basilides offered a principle via which we can achieve the desirable balance between these two seemingly conflicting states of being. As its name implies, *Principium Individuationis* is the safeguard of the individuality (aka distinctiveness) of human nature. Differentiation of our being leads to a return to our true nature. This newly offered principle could rescue temporarily-bound humanity from the dangerously infinite Pleromatic state.

Humans' individuality, which materializes through pairs of opposites, is vitally connected to temporality. Any loss of this connection endangers the species' nature. The realm of the Pleroma as presented in this poem offers a unique combination of a person within a time-and-space-limited context, accounting for timelessness. In *Septem Sermones* Basilides presented the manner in which individuals in time can incorporate their timelessness; in the poem the self (presented as Pleroma) was viewed by Jung in a *temporal perspective* for the first time.

After Basilides explored Pleroma and Creatura, he dealt with various entities found within these realms. In the Second Sermon he spoke of the God Helios (sun) who, like humans, is distinct from Pleroma and belongs to Creatura. However, Helios is more indefinite and indeterminable; he is the effective fullness of the Pleroma and brings forth generation. In opposition to Helios stands the effective emptiness, the source of destruction, the Devil. Like all other qualities perceived by created beings, 'God and Devil … do not cancel each other out, rather they stand against each other as active opposites' (Jung, as cited in Hoeller 1982: 49). Action characterizes both God and the Devil: 'Activity is common in both. Activity unites them. It is for this reason that activity stands above both, being God above God, for it unites fullness and emptiness in its working' (Jung, as cited in Hoeller 1982: 50).

Basilides introduced yet another being identified by activity: the god Abraxas. This god is unknown to humans as he stands above the god Helios and his counterpart the Devil. Even though the God and the Devil appear more powerful since they have definite effects, this is a misconception. Abraxas is far greater than all other gods, but since he is indefinable we cannot directly observe his almighty force. Abraxas is force, duration and change.

Abraxas, according to the Gnostic system, is a rooster-headed god with serpent feet. It is believed that his name derived from numerological principles and corresponds to 365, the number of days in a calendar year.[19] Abraxas is identified as:

> The deity that rules over the totality of time and by whose power time is both made and unmade. The unmaking of time ... is a true Gnostic concern. In a sense it might be regarded as the hallmark of the true Gnosis, possessed only by the most highly accomplished pneumatic or spiritual individual. Abraxas, as the all-pervading energy of being, is thus the sum of and the liberator from the cycle of necessity, freeing man from the agony of time ... [Abraxas is] the eternally available timeless moment, the eternal now, the recognition and utilization of which brings freedom from time both in its linear and its cyclic aspects.
>
> (Hoeller 1982: 87)

For the Gnostics the ultimate goal is the return to the Pleromatic state, which is considered to be a timeless/atemporal dimension. The object of salvation is to deliver us from the lie of time: 'Salvation places us outside time ... it releases us from time in order to carry us back to our original atemporal condition' (Puech 1957: 77). Abraxas, then, is the deity who eliminates the obstacle of time from the soul, the one who lifts and surpasses all time-related barriers and limitations.

Jung shared with the Gnostics the belief that tension between opposites must exist for the generation of vital energy. The synergetic existence of pairs of opposites is fundamental for the libido. Life must be experienced in time as well as in connection to the eternal and timeless. Similarly to libido, Abraxas is the source of life/energy.[20] Abraxas is the ruler, maker and undoer of time that holds the temporal tension.[21]

It has also been theorized that Abraxas can be viewed as a symbol of the archetype of the Self.[22] Both are representations of the ultimate wholeness from which all opposites stream.

Abraxas cannot be experienced directly but rather through various emanations including good and evil. Psychic reality, according to Jung's later writings, includes the shadow side as well as the positive pole, both of which are united in the majestic self. As Basilides emphasized in the

Forth Sermon, it is imperative to accept multiplicity rather than a single one-sided godhead:

> Blessed am I, for it is granted unto me to know the multiplicity and diversity of the gods. Woe unto you, for you have substituted the oneness of God for the diversity which cannot be resolved into the one … How can you be true to your nature when you attempt to make one out of the many?
>
> (Jung, as cited in Hoeller 1982: 53)

Gods and demons are equally vital and essential in the realization of the self. As with all other pairs of opposites, gods and demons are interrelated syzygies, which are equally responsible for the attainment of wholeness.

Taking the synergetic function of the pairs of opposites a step further, in *Septem Sermones* time and its counterpart eternity can be seen as such a pair – a pair that converges and interplays in cosmic and individual planes. The mortal human does not function in exclusion of eternity; eternity is incorporated into the time-bound human existence: 'wholeness must be lived in time and must also be experienced in timelessness' (Hoeller 1982: 151). In Abraxas, the bird-headed serpent, time-bound humans join their eternal counterpart.

Contrary to the one-time event of salvation in the Christian religion (the Second Coming or Judgement Day) for the Gnostic tradition liberation is a repeatable event in the present.[23] Thus, in Gnosticism the linear progression of time as defined by Christianity is transformed and with it the anguish conveyed by time is abolished. The atemporal and eternal power of Abraxas is the key to the soul's deliverance, which can be obtained repetitively at the present moment. When time is tamed and ruled over, psychic salvation is attainable.

Once Basilides had explored the world of opposites with their tension and unification, the dead came back in the final sermon and asked to be enlightened 'concerning man'. Here Basilides portrayed man as a link between the macrocosm and the microcosm, a doorway where the two worlds meet:

> Man is a portal through which one enters from the outer world of the gods, demons and souls, into the inner world, from the greater world into the smaller world. Small and insignificant is man; one leaves him soon behind, and thus one enters once more into infinite space, into the microcosm, into the inner eternity.
>
> (Jung, as cited in Hoeller 1982: 57–8)

Basilides presented the manner by which mortal time-bound humans can integrate their eternal nature. In man the cosmic and the intrapsychic meet; in his psychic reality the outer world joins the inner realm. The seemingly interior world and the seemingly exterior reality meet in 'infinite space' and 'eternity'.

Hoeller (1982) suggested that the above description of a meeting between the outer and the inner world is related to Jung's synchronicity. The world of 'gods, demons and souls' could be seen as the realm of archetypes, which are projected onto the macrocosm of sense-perception. Jung described a synchronistic experience in a similar manner: as the coming into the foreground of an archetypal experience. During synchronistic occurrences the personal and the archetypal realms unify. In Chapter 3 the principle of synchronicity will be presented in detail. Nonetheless, it is worth mentioning at this point that decades after the composition of the poem Jung included the Gnostic terminology in his account:

> What exists in the pleroma as an eternal process appears in time as an aperiodic sequence, that is to say, it is repeated many times in an irregular pattern ... When these things occur as modern variants, therefore, they should not be regarded merely as personal episodes, moods, or chance idiosyncrasies in people, but as fragments of the pleromatic process itself, which broke up into individual events occurring in time, is an essential component or aspect of the divine drama.
>
> (Jung 1952a: par. 629)

There is temporal opposition between an eternal process and occurrences in time. During a synchronistic experience fragments of the Pleromatic eternity enter the time-bound realm; events happen in time. Simultaneously there is a divine aspect to them. In the years following *Septem Sermones* Jung refined the dynamics introduced in the Gnostic poem. Time's role was of particular importance to these elaborations; its function, as first presented in the poem, would later provide the foundation for key elements of his theory.

Notes

1 'It's a mistake to think it's the small things that we control and not the large, it's the other way around! We can't stop the small accident, the tiny detail that conspires into fate: the extra moment you run back for something forgotten, a moment that saves you from an accident – or causes one. But we can assert the largest order ... the only order large enough to see' (Michaels 1997: 22).

2 Milton 1905: 190.

3 For Jung the libido is 'the unconscious creative force which wraps itself in images' and 'whose nature we can only experience through its effect on us' (Jung 1911–12/1952: par. 329). Jung clarified that for him libido does not have the sexual connotation that it does for Freud, but he nonetheless chose to use the same word 'if only for reasons of historical justice' (Jung 1928a: par. 54).

4 Jung first examined the symbol of the horse, which he linked with numerous mother-symbols, such as the Terrible Mother who becomes the underworld and leads one into devilish temptations (Jung 1911–12/1952: par. 315), as well as the projection of the mother symbol upon water (ibid.: pars. 319, 320). The common factor in the mother and the horse, as well as in various other symbols presented (such as the lion and the snake) is the libido: 'The mother-imago is a libido-symbol and so is the horse; at some point the meaning of the two symbols overlaps. But the factor common to both is the libido' (ibid.: par. 421).

5 Basilides (AD 117–138) was a prolific and influential scholar who taught in Alexandria during the reign of Emperor Hadrian (Rudolph 1987; Schneemelcher 1991). In *The Red Book* it is not Basilides, but rather Philemon who is in conversation with the dead. Jung also identified Philemon with the Gnostic prophet Simon Magus (Jung 2009).

6 During his lifetime Jung allowed only a few members of his family and friends to read *The Red Book*. In October 2009 the book was published for the first time (Jung 2009).

7 'Both Marie-Louise von Franz and Barbara Hannah consider this piece an error and clarify that Jung regretted the publication of the book, even in its private edition, but not its contents' (Papadopoulos 1980: 291).

8 It is useful to briefly review the nature of Gnosticism so as to better understand its influence on Jung's writings. It is perhaps easier to describe what Gnosticism is not. In the Gnostic tradition there was no church, theology, rule of faith, dogma or canon of scripture. The Gnostics used freely a variety of religious and mythological sources for the purpose of sanctioning their own teachings. Thus, various otherwise incompatible traditions (such as Christian theology and Homeric mythology) can be found side by side in Gnostic scriptures. The Gnostics would use this unconventional epistemology – known as the method of *exegesis* – in order to derive their own meaning, ideology and cosmology. This unorthodox tolerance and utilization of diverse sources is perhaps one of the reasons why the Gnostic tradition is difficult to understand and interpret. Nonetheless, the Gnostic system has a sense of coherence made possible by the *Gnostic myth*. Central to this myth is *Gnosis*, i.e. the immediate esoteric knowledge of the spiritual truth, given by the divine to the selected few. The essential idea of *Gnosis* rests on the existence within man of a divine spark; man was initially part of the divine world but has since fallen into the material world of destiny, birth and death. The goal of man is to reawaken his divine counterpart (with the attainment of *Gnosis*) and thus be restored, redeemed and liberated. For an elaboration of Gnosticism's premises, see Hoeller 1982; Rudolph 1987; Dehing 1990.

9 'By attributing this poem to Basilides, Jung, in a sense, disclaimed the authorship of its contents by distancing himself in order to afford a clearer and more objective study of the workings of his psyche' (Papadopoulos 1980: 303).

10 Hubback 1966; Hoeller 1982; Brenner 1990; Dehing 1990; Main 2004.

11 'Jung as Basilides' will henceforth be referred to as Basilides.
12 The word Pleroma comes from the Greek *πλήρωμα*, and it suggests a fulfilment of purpose.
13 Puech 1957.
14 For the purposes of this chapter, I used Stephan Hoeller's translation of *Septem Sermones Ad Mortuos* as found in his book *The Gnostic Jung and the Seven Sermons to the Dead* (Hoeller 1982: 44–58).
15 The word 'Creatura' appears to be a derivative of the Latin '*creare*' (as is the English verb 'to create', to cause to exist; to bring into being).
16 Jung 1935: par. 140.
17 The etymology of the word 'syzygy' implies the unification between two distinct but similar parts: 'syzygy' (*συζυγία*) derives from the composites *συν* meaning together, and *ζυγός* which means yoke.
18 Hubback 1966; Hoeller 1982; Brenner 1990.
19 Brenner 1990; Hoeller 1982.
20 Papadopoulos 1980.
21 An indirect connection of Abraxas to time can be found in Jung's early writings. Abraxas was often depicted as mounted on a chariot with four white horses representing the four forces by which the psychic energy is expressed, as well as the four elements (earth, water, fire and air). Jung referred to the symbolism of the four-horse chariot as a symbol of time (Jung 1911–12/1952: pars. 423–5).
22 Hoeller 1982.
23 Ibid.

Chapter 2

The unfolding of Jung's time theory

Dreaming the myth onwards

> Who would fare better in this world of fitful time?
> Those who have seen the future and live only one life?
> Or those who have not seen the future and wait to live life?
> Or those who deny the future and live two lives?
>
> Alan Lightman, *Einstein's Dreams: A Novel*[1]

In the previous chapter I examined the vital role of time, in the context of *Septem Sermones* for the distinction between two realms: time defines Creatura; mortal humans who belong to this sphere are time-bound. In contrast, Pleroma is timeless and eternal. At the time of writing *Septem Sermones* Jung's theory was still quite amorphous; the poem served as a precursor of his later psychological understanding. In the years following the writing of the Gnostic poem Jung would apply the temporal complexities and implications of the realms of Pleroma and Creatura in his theory. The opus of his work following the poem was in effect an elaboration of the implications first presented in a poetic language.

The way Jung understood and developed fundamental aspects in his theory, such as the relation between conscious-unconscious, can be seen as an amplification of Basilides' cosmology. In the realm of Creatura differentiation is a requirement for consciousness: 'The whole essence of consciousness', Jung wrote, 'is *discrimination*, distinguishing ego from non-ego, subject from object, positive from negative, and so forth. The separation into pairs of opposites is entirely due to conscious differentiation' (Jung 1921: par. 179). Likewise, similar to Basilides' definition of the Pleromatic realm, the unconscious is characterized by indiscrimination and lack of differentiation: 'Where no consciousness exists, where purely unconscious instinctive life still prevails, there is no reflection, no *pro et contra*, no disunion' (ibid.). Discrimination is '[t]he hallmark of the conscious mind ... it must, if it is to be aware of things, separate the opposites, for in nature the opposites merge with one another. Yet, once separated, the two must be consciously related to one another' (Samuels *et al.* 2003: 37). These descriptions are reminiscent of Basilides' words: humanity belongs to the realm of Creatura where discrimination reigns; separation of opposites is a necessity for the human condition which, nonetheless, must also come to terms with its Pleromatic and non-differentiated nature.

Four decades after the poem was written Jung would revisit the world narrated by Basilides in order to elaborate and integrate the role of time in Creatura and Pleroma in a more systematic manner. In Gnostic language he explained that the Pleroma was the realm of knowledge (*gnosis*) where a perfect interplay existed between the cosmic forces. This pleromatic state was disrupted by Creation. Creation is identical to the establishment of time and space in the universe. Time is an element of a 'less than perfect' state of the world where there are limitations and chasms between the original state of things and the earthly: 'In the pleromatic ... state, there is a perfect interplay of cosmic forces, but with the Creation – that is with the division of the world into distinct processes in space and time – events begin to rub and jostle one another' (Jung 1952a: par. 620).

The new order, of which temporality was a defining element, brought about imperfection since events were henceforth in contact with each other in time and space. Time not only differentiated the two realms but was responsible for the initial split between the two. Once time was established, imperfection was introduced in the universe. From that moment onwards there were limitations to cosmic reality, its oneness was shattered and its shadow was generated. The birth of time and the birth of humanity occurred concurrently. With time came change, differentiation and consequently degeneration and mortality, which disturbed the Pleromatic state of the universe.

Jung's understanding of temporality matured after the writing of the poem. Beyond the occasional utilization of the Gnostic style in his later writings, *Septem Sermones*' imagistic language assisted in various aspects of his psychology and their relation to the temporal element. An investigation of his post-*Septem Sermones* writings, with particular attention paid to the temporal parameter, reveals certain important implications for his overall theory. Time can provide an illuminating factor through which key elements in his theory can be examined. I would even argue that, the common thread in Jung's conceptualizations is an underlying theory of temporality in the psyche. His 'time theory' was the stage upon which his psychology unfolded.

The paradoxical temporality of the archetype

The archetype is the aspect of Jung's theory most blatantly associated with temporality. In descriptions of its qualities there are several contradictory mentions of time and its lack. Jung suggested that 'the archetypal world is "eternal," i.e., outside time' (Jung 1976: 46). Here eternal seems to indicate a condition that does not follow the laws of time. When denoting the inherent qualities of the archetypes he commented upon their independence of time. He proposed that archetypes have a timeless quality that is 'beyond individual birth and death' and therefore posses 'an "eternal" presence' (Jung 1936: par. 329). By consciously integrating the material of the unconscious it is possible to 'make experiences of an archetypal nature providing us with the feeling of continuity before and after our existence. The better we understand the archetype, the more we participate in its life and the more we realize its eternity or timelessness' (Jung 1954: par. 1572). Jung invited his readers to explore in depth the nature of archetypes that rests on a temporality beyond a personal sense of time.

Contrary to the eternal and timeless realm of the archetypes, 'the individual is continuously "historical" because strictly time-bound' (Jung 1938/1940: pars. 146–8). Plainly stated, 'the psychic life of the archetype is timeless, in comparison with our individual time-boundness' (ibid.: par. 149). However, the archetypal realm – which is the ground for all psychic activity – and the individual life are 'so much interwoven ... that in every case considerable critical effort is needed to separate the unique from the typical with any certainty' (ibid.: pars. 146–8). While there is discrepancy between the temporality – or lack thereof – of the archetypal and the personal, the individual life also belongs to the timeless and eternal realm of the archetypes so that, 'Ultimately, every individual life is at the same time the eternal life of the species' (ibid.).

In Jung's understanding of the relation of the two realms (the archetypal and the personal) with the element of time, we come across the same paradox introduced in *Septem Sermones*. Humans, who are defined by time, are simultaneously part of the Pleromatic and eternal realm. We have a *dual temporal nature*. We are time-bound and function in time but also have an archetypically timeless dimension which allows us to participate in a realm that is beyond/outside time, and thus timeless. The individual mortality-bound life is simultaneously connected to an eternal aspect of the collective.

There is another paradox in the defining qualities of the archetypes, and consequently in the collective unconscious which houses them. On the one hand the archetypes are timeless and belong to the eternal. On the other, they have a definite temporal quality, since they develop through generations:

> The collective unconscious, being the repository of man's experience and at the same time the prior condition of this experience, is an image of the world that has taken aeons to form. In this image certain features, the archetypes or dominants, have crystallized out in the course of time.
>
> (Jung 1917/1926/1943: par. 151)

Archetypes are primordial and archaic and 'have existed since the remotest times' (Jung 1934/1954: par. 5); they are the result of 'the psychic life of our ancestors right back to the earliest beginnings' (Jung 1929: par. 230). During the passage of time life events form certain patterns of experience and behaviour. Similar to the human psyche, while archetypes have a timeless and eternal quality they have, nonetheless, developed *through time* and thus also belong to the temporal realm.

Retrospection and forward movement

Archetypes have a beginning in the ancestral past as well as a future orientation. This *temporal bidirectionality* on time's spectrum is one of the key differences between Freud's and Jung's perspectives. For Freud the aim of psychoanalysis was to *re*-construct; to recreate the past, to revisit it. The past is the point of departure as well as the focal point along the timeline.[2] In comparison to Jung, it appears that there is a disproportionate attention paid to the past by Freud and an absence of the future dimension.[3]

Freud's and Jung's opposing approaches regarding the future becomes evident via their respective views on future-telling dreams. According to Freud, a key temporal aspect of the dream is its orientation towards the past. In the concluding paragraph of *The Interpretation of Dreams* Freud blatantly rejected the concept of prophetic dreams, theorizing instead that these kinds of dreams – similar to all dreams – aim at wish-fulfilment.

> And the value of dreams for giving us knowledge of the future? There is of course no question of that. It would be truer to say instead that they give us knowledge of the past. For dreams are derived from the past in every sense. Nevertheless the ancient belief that dreams foretell the future is not wholly devoid of truth. By picturing our wishes as fulfilled, dreams are after all leading us into the future. But this future, which the dreamer pictures as the present, has been moulded by his indestructible wish into a perfect likeness of the past.
>
> (Freud 1900: 621)

There is no room for the future in its own right; it is an extension, a replica even, of the past. The future-telling aspect of the dream is a mere projection of the wish. The seemingly predictive quality of dreams is the result of repetition of the past, which creates 'a template of the future' (Perelberg 2008: 34). Freud's theory presents a temporal loophole, a vicious cycle where the past is bound to repeat itself and the future is viewed as a mirror of what has already taken place. The focus remains on the past, on its reinterpretation, repetition and rediscovery.

In order to clarify his view on premonition dreams, in the first Appendix of *The Interpretation of Dreams* Freud presented the case of Frau B., a patient who claimed to have dreamed of an encounter with someone the night before it had actually occurred: 'On seeing [the man in Frau B.'s dream] she felt convinced that she had dreamt the night before of having this very meeting at that precise spot' (Freud 1900: 623). Freud rejected the idea that this was a premonition dream with a dose of irony. The dream was a derivative of a current thought which was suppressed upon waking.

> And that was what happened to our ostensibly prophetic dream ... Thus the creation of a dream after the event, which alone makes prophetic dreams possible, is nothing other than a form of censoring, thanks to which the dream is able to make its way through into consciousness.
>
> (Freud 1900: 625)

The dreamer herself characterized a similar occurrence – during which a friend appeared at her doorstep while she was crying 'in passionate longing' for him – as 'a remarkable coincidence' (ibid.). Freud disputed her account: 'We shall find nothing at all remarkable in this coincidence when we consider how often she thought of him and how often he probably visited her' (ibid.: 624). He seemed to have felt very strongly against the possibility of 'a remarkable coincidence', and thus went beyond the reported facts to presume that the meeting must have been unremarkable.

The dream example, as well as the above incident that Freud used in order to dispute prophetic dreams, have a striking resemblance to some of the instances cited by Jung as illustrations of the principle of synchronicity. For Freud, the fact that the dreamer reported reliving in waking life a scene that she had already experienced in a dream was an example of censoring of the unconscious. For Jung this would most likely have been a suitable illustration of an internal event repeating externally, and thus it would constitute a synchronistic occurrence.

Even though Jung emphasized the future in his theory it is inaccurate to suggest that he downplayed the importance of the past. On the contrary, his concept of the collective unconscious can be viewed as an emphasis towards the collective past. The past forms the roots to consciousness. At each new stage of our development it is essential, for the sake of our mental health, to 'orient to the past' in order to reconnect with our inherited origins.

> In reality we can never legitimately cut loose from our archetypal foundations unless we are prepared to pay the price of a neurosis ... If we cannot deny the archetypes or otherwise neutralize them, we are confronted, at every new stage in the differentiation of consciousness to which civilization attains, with the task of finding a new interpretation appropriate to this stage, in order to connect the life of the past that still exists in us with the life of the present, which threatens to slip away from it. If this link-up does not take place, a kind of rootless consciousness comes into being no longer oriented to the past, a consciousness which succumbs helplessly to all manner of suggestions, and in practice, is susceptible to psychic epidemics.
>
> (Jung 1940: par. 267)

Jung was unambiguous about the need to stay connected to our past, not to deny it, and to continually update our connection with it, by reinterpreting it.

For Jung, the past does not imply a particular moment in history but rather an earlier epoch; it refers to a *past dimension* of the psyche, rather than a particular past event or moment. This is why Jung's interpretation of (what Freud labelled) phylogeny is not located at a particular point in history:

> [F]antasies (including dreams) of an impersonal character, which cannot be reduced to experiences in the individual's past, and thus cannot be explained as something individually acquired ... correspond to certain *collective* (and not personal) structural elements of the human psyche in general, and, like the morphological elements of the human body, are *inherited.*
>
> (Jung 1940: par. 262)

Also, revisiting the past goes beyond pathology. The connection and reinterpretation with the past is vital for our everyday healthy functioning. Jung emphasized the modernization of the past. We need to not simply revisit the past but to reinvent it through our current standpoint, to see it not as it was, but as we understand it in the present.

> Our myth has become mute, and gives no answers. The fault lies not in it as it is set down in the Scriptures, but solely in us, who have not developed it further, who, rather, have suppressed any such attempts.
>
> (Jung 1961: 332)

Constant development is required for the voice of the myth to be heard, for answers to be given through it.

Archetypes themselves are dynamic since they do not simply exist in the distant past, but are active in the present. Their function, that is, their teleological significance, is to give meaning in the present by connecting the 'now' with the past. This is illustrated through the nature of the child archetype.

> The child motif represents not only something that existed in the distant past but also something that exists *now*; that is to say, it is not just a vestige but a system functioning in the present whose purpose is to compensate or correct, in a meaningful manner, the inevitable one-sidedness and extravagances of the conscious mind.
>
> (Jung 1940: par. 276)

Archetypes are the catalysts that transport our psyche from the past into the present, and move it into the future. In essence, if we manage to stay connected with our roots we will be carried successfully *towards the future*. This is an essential dissimilarity between Freud's and Jung's temporal spectra. For Jung, the past is vital, but of equal importance is the future.

> One of the essential features of the child motif is its futurity. The child is potential future. Hence the occurrence of the child motif in the psychology of the individual signifies as a rule an anticipation of the future developments, even though at first sight it may seem like a retrospective configuration.
>
> (Jung 1940: par. 278)

What is to all appearances a turn backwards into the past is, in fact, a forward movement into the future.

Jung clearly stated his belief regarding the temporal directionality of the psyche: 'Life is a flux, a flowing into the future, and not a stoppage or a backwash' (ibid.). There is dynamic, purposeful movement forward. Retrospection is necessary so that the psyche can stay rooted in the past and thus travel forward, towards its *telos* of self-realization. The anchoring in the past dimension assures a forward movement. Thus, not only the entire temporal spectrum is taken into account but it is considered vital in the process of individuation, which has as its goal 'the synthesis of the self'[4] (ibid.). Wholeness is achieved by the unification of the opposites that are embedded in the archetype. The child is *simultaneously* the past and the potential future.

Jung's belief in the teleological aim of the psyche clashed with Freud's view regarding the retrospective directionality of the psyche. According to Freud's early postulations, consciousness resists remembering the original trauma, which is the cause of the symptom. Jung's disagreement with Freud is evident in his early writings. For Jung the actual event in the past is not as significant as the fantasy of the event. If we try to trace the trauma back to its original source, in the historical past, we follow the causal path that leads to greater resistance. Instead we must understand resistances 'not as defences against the conscious realization of a painful conscious memory, but as a struggle against the construction of this fantasy' (Jung 1913: par. 391). A fantasy is formed at a later point in time based on the original event; this fantasy is what consciousness resists. The libido regresses to 'a secondary fantastic dramatization' (ibid.: par. 403) of the original traumatic event. Consequently, when we follow the regression (i.e. the libido's retreat into the past) we are not following historic developments, but 'a subsequently formed fantasy based only in part on former realities' (ibid.: par. 394).

Jung suggested that analysts, implying Freud, seek the origins of fantasies in the distant past due to their 'need for causality' (ibid.: par. 391). This orientation is futile.

> The further we get away, in analytical investigations, from the epoch of the manifest neurosis, the less can we expect to find the real *causa efficiens*, since the dynamics of the maladjustment grow fainter and fainter the further we go back into the past ... *For the cause of the pathogenic conflict lies mainly in the present moment.*
>
> (ibid.: par. 373; original italics)

The past is a safer time than the present moment, and thus the libido regresses there, creating a fantasy around that time. It is, however, only that: a fantasy. And since 'the fantasy material brought out by analysis became significant only afterwards, we are not in a position to use this material to explain the onset of the neurosis; we should be constantly moving in a circle' (ibid.: par. 395). It is not the actual event that is important, but rather the *regressive fantasy* that signifies the 'moment' that the event became meaningful.

Regressive tendencies do not just direct the psyche into the past. On the contrary, they have a teleological value: '[The neurotic's] retreat to the infantile level does not mean only regression and stagnation, but also the possibility of discovering a new life-plan. Regression is thus in very truth the basic condition for the act of creation' (ibid.: par. 406). Jung's temporal directionality is re-emphasized: even when regression takes place, it happens for the sake of forward movement into an improved future. The teleological value of the regression is preparation for the creative movement forward, towards a new life.

It has been indicated that Jung's theory regarding regressive fantasies greatly influenced Freud's understanding and has instigated a revision of his (Freud's) theory of the primal scene.[5] In the 'Wolf Man Case Study' Freud (1918) theorized that a dream that had occurred twenty years later gave rise to his patient's neurosis because he was able to grasp retrospectively the meaning of his past experience; the past events were reinterpreted based on current situations. However, in 1917 Freud returned to the case study to apply his revised theory, stating that it is possible that the primal scene was constructed during analysis, that is, the memory of it might have been manufactured by the patient at a later point in time. In Freud's corrected version the primal scene need *not* be a memory based on actual events.

According to this account, this disagreement was at the centre of the storm that broke between Jung and Freud. Jung's belief that reconstruction was invented in retrospect clashed with Freud's original view that the *actual past event* caused pathology. Freud was unsettled by Jung's theory that fantasies occurring at the present moment act retroactively and shape the past and thus felt the need to revise his original thesis by proposing that the traumatic event, even though it did not necessarily take place in real life, nonetheless exists as a psychic reality.

The abandonment of the belief that an actual sexual trauma that occurred in the past had caused neurotic symptoms marked a significant shift in Freud's theory. However, even though he modified his original views, perception (i.e. actual observations, based on one's senses) was still, at least partially, responsible for the psychic experience. The based-on-psychic-reality-memory of the primal scene, even though not a memory of an actual event, was believed to have caused retrospective meaning-making.

After his break with Freud and as his thought matured, Jung was unambiguous about the role of regression and the teleological direction of the psyche. Freud's approach as Jung understood it was primarily retrospective in that it prioritized the uncovering of past events that have caused the present issue. Even though Jung recognized the importance, and even the inevitability, of retrospection – 'my criticism [of Freud] would not go so far as to deny the extraordinary power of the retrospective tendency. On the contrary, I consider it to be of the greatest importance' (Jung 1930: par. 54) – he was opposed to the idea that the return to a previous stage is the most desirable movement for the psyche and the solution to the issue at hand. Jung unequivocally considered the future as more advanced than the past. Movement forward is what Jung aspired to.

> [There exists] a belief in the illusory picture of an earlier, more perfect type of man, who somehow fell from his high estate. Retrospective orientation is itself a relic of pagan thinking, for it is a well-known characteristic of the archaic and barbarian mentality that it imagined a paradisal Golden Age as the forerunner of the present evil times. It was the great social and spiritual achievement of Christianity that first gave man hope for the future, and promised him some possibility of realizing of his ideals.
>
> (Jung 1921: par. 124)

Jung rejected the notion that the past holds a more secure state for the psyche – 'has it ever been thought that the golden age of Paradise lay in the past!' (Jung 1913: par. 395). The psyche looks hopefully ahead into the future, gazing towards a more developed stage.

Like his interpretation of the past, for Jung the future does not signify a point in time down the linear temporal line (i.e. a particular future occasion). The future is *a dynamic dimension* that influences the past and the present.

> From another point of view the term 'entelechy' might be preferable to 'synthesis' [in describing the goal of individuation]. There is an empirical reason why 'entelechy' is, in certain conditions, more fitting: the symbols of wholeness frequently occur at the beginning of the individuation process, indeed they can often be observed in the first dreams of early infancy. This observation says much for the *a priori* existence of potential wholeness, and on this account the idea of *entelechy* instantly recommends itself.
>
> (Jung 1940: par. 278)

Jung's attention to semantic details could be criticized as trivial, but in fact his distinction is important in understanding the temporal foundations of his theory. In his 1936–40 dream seminars (Jung 2008), he elaborated on his belief that the potential personality that could exist at a future point in time is present in a child's dreams. For that reason, the choice of the word 'entelechy' is indeed accurate; it is ideal when we take into account the influence of *potentials* existing since the beginning of a lifetime/process. Entelechy deals with the actualization of *telos* (embedded in its etymology). It is the vital force that directs an organism towards self-fulfilment. In fact, the word itself has an inherent teleological force. Indeed, when referring to the potential wholeness that exists *a priori* in the psyche, 'on this account the idea of entelechy instantly recommends itself' (Jung 1940: par. 278); it is as if the word was created to explain the process described by Jung.

Perhaps, then, we can better understand the disagreement between Freud and Jung regarding future-telling dreams. For Freud foretelling the future is an absurd impossibility. There is no way of travelling forwards on the linear timeline and seeing what has not occurred yet. For Jung a possible future is contained in the present and past, and is, therefore, not only a probable aspect of the dream but an essential one. The future and the past dimensions are dynamically active *at all times*.

Perelberg noted that Breton [1969] in *Manifestoes of Surrealism* 'suggested that Freud had been mistaken in not believing in the prophetic nature of dreams ... As a consequence he denied dreams the quality of movement' (Perelberg 2008: 45). Perhaps it is more accurate to suggest that, by rejecting the possibility of the dream's future-telling ability, Freud denied the dream the quality of *forward movement*, and thus restricted the overall temporal scope of the psyche.

Meaning cannot be restricted to the past. A vital aspect of meaning is its potentialities, which point towards the future.

> Meaning has implications and, therefore, consequences. Meaning cannot avoid putting itself into the future (as it does into the past) because it is a dynamic agency; and, because its movement comes within the transforming potentialities of what is possible, whose future and past have to be invented.
> (Green 2002: 42–3)

If meaning is to be given a rightful place, the scope of the psyche cannot be limited to a backward movement – characterized by the 're' prefix – but should include dynamic forward movement.

Teleology and causality in the psyche: methodological considerations

The archetype's future orientation was emphasized throughout Jung's work. In his accounts of goal-orientation and foreknowledge its future dimension is unambiguous. The archetype determines not only the nature of the emotion but also 'the course it will follow, with seeming foreknowledge, or as though it were already in possession of the goal to be circumscribed by the centring process' (Jung 1947/1954: par. 411). Not only the contents of the archetype operate in time and thus have a temporal existence, but also *the whole of the psyche*.

> On the one hand [the psyche] gives the picture of the remnants and traces of all that has been, and on the other, but expressed in the same picture, the outline of what is to come, in so far as the psyche creates its own future. The psyche at any given moment is on the one hand the result and culmination of all that has been and on the other a symbolic expression of all that is to be.
>
> (Jung 1914: pars. 404–5)

How can we explore this all-inclusive aspect of the psyche so as to embrace its historic past, its present, as well as its symbolic future? To address this problem Jung had to steer away from the traditional approach widely employed by the great majority of his contemporary theorists and adopt a different methodology.

The scientific community since the beginning of the seventeenth century adhered to determinism, which supported the cause-and-effect explanation of events. Science consisted of 'discovering the frame and operations of nature, and deducing them, as far as may be, to general rules or laws – establishing these rules by observations and experiments, and thence deducing the causes and effects of things' (Newton 1952: 3).

Jung rejected the exclusive use of this method and suggested instead a different way of explaining the arrangement of events. He asked his reader to 'remember that the rationalistic attitude of the West is not the only possible one and is not all-embracing, but is in many ways a prejudice and a bias that ought perhaps to be corrected' (Jung 1952c: par. 916). He advocated an approach that would allow for the exploration of the psyche's totality.

> [M]odern science is identical with the *causal principle*. As soon as we regard the psyche causally, that is scientifically, the psyche as a creative function eludes us. If we want to grasp the other side of the psyche, we shall never do it by the exclusive application of the causal principle, but only with the help of the constructive standpoint. The causal standpoint reduces things to their elements, the constructive standpoint elaborates them into something higher and more complicated.
>
> (Jung 1914: pars. 404–5)

Since his earlier work Jung supported the view that understanding the totality of the psyche – which extends into the past as well as the future – requires the employment of an additional principle. In an Aristotelian fashion, he believed that beyond the efficient cause that pushes a process into action there is a final cause, which explains processes by means of their final expression.

Albeit technically inaccurate and somewhat vague, Jung associated efficient cause with causality (as well as the causal principle, the causal view and the causal standpoint) and final cause with teleology (as well as finality and the constructive standpoint).[6] He explained that (what he labelled as) causality, overlooks the 'final significance' of events, whereas, 'according to the concept of *finality* causes are understood as means to an end' (Jung 1928a: pars. 43–4)[7], thus emphasizing the significance of an event's function and future direction. He advocated the use of finality, stating that 'it is impossible to consider the psyche from the causal standpoint only; we are obliged to consider it also from a final point of view' (Jung 1916/1917: par. 688) To approach the psyche, as well as the physical world, from a symbolic perspective the teleological perspective must be adopted because:

> [W]hat to the causal view is *fact* to the final view is *symbol* ... A symbolic interpretation of causes by means of the energic standpoint is necessary for the differentiation of the psyche, since unless the facts are symbolically interpreted, the causes remain immutable substances which go on operating continuously, as the case of Freud's old trauma theory. Cause alone does not make development possible.
>
> (Jung 1928a: pars. 45–6)

The teleological standpoint allows for the exploration of the psyche's totality, its past, present and future. And since the future phase of the psyche's development has not occurred yet, we must utilize symbolic language (Jung 1921). By treating the products of the unconscious as symbols, rather than concrete outcomes, the teleological approach allows for various possible future incarnations of a certain archetypal theme.

Since the future is not a concrete point down a linear timeline *telos* is not perceived as a literal end-of-the-line state. By emphasizing the symbolic aspect, Jung moved away from causality-driven explanations and opened the way to an abstract future where events are yet to solidify. There is a significant distinction between a symbolic versus a factual *telos*. As Hillman emphasized, we must not literalize *telos* by assuming a perceivable and concrete 'goal' existing *a priori* somewhere in the actual future. If we mistakenly literalize the concept as intended by Jung's teleological approach, 'then we move the *telos* from an internal component to an external image; then we take it literally, placing futurity in the future, a mistake as naïve as taking the efficient cause literally and placing it in the past' (Hillman 1975: 34)

Hillman's position contradicts Freud's theory which maintains that a particular past event (personal or historical) is the cause of the present trauma. It also supports the idea that for Jung past and future are perceived as dimensions, rather than temporal demarcations on a timeline. *Telos* is an image that encompasses futurity and hope without being specific or concrete. The *telos* implied in final cause is *not an eschatological viewpoint*, just as there is no particular origin implied in efficient cause. If we talk literally of the psyche's goal, we '[abort] it of its teleological impulse' (ibid.: 35).

If we are to understand Jung's teleology an imagistic viewpoint must be employed where the symbolic dimensions of the future, the past and the present provide the psyche with a dynamic movement forward. The psyche, due to its teleological impulse, is inclined towards futurity; but not towards a concrete future defined by temporal markers. *Telos* is not based on linear temporality, but rather it is activated *now*. Not now as in 'the present moment', but now as the force that is constantly present and gives vitality and movement to the psyche.

Jung criticized Freud's causal viewpoint and stated that if we adopt such an approach we see things 'as mere historical survivals and metastases of the incest taboo ... If, however, we approach the problem from the teleological point of view, much that was quite inexplicable becomes clear' (Jung 1928b: par. 239). It has been observed that Jung's and Freud's 'respective inabilities to tolerate deterministic and teleological explanations were a major determinant of their eventual break' (Horne *et al.* 2000: 111).

The future directionality of the psyche is what Jung attempted to capture by the adoption of teleology as an additional principle to causality. Teleology implies future-oriented movement to a point in time where (ideally) a more improved and integrated version of the psyche lies. Growth awaits in the future. Even in the event of regression, which implies a blatant movement backwards, there is a teleological, future-directed point of development. For Jung, life is an energy-process with an obvious teleological urge: 'Life is teleology *par excellence*; it is the intrinsic striving towards a goal, and the living organism is a system of directed aims which seem to fulfil themselves. The end of every process is its goal' (Jung 1934: par. 798). The teleological approach adds a new dimension to the strict causal methodology and brings forth a new level of clarity.

Teleology and causality are two distinct methods of interpretation, two opposing approaches for scientific understanding. The principle of teleology should be employed since it is beyond causality and also includes the psyche's symbolic function. However, even though Jung favoured the teleological approach he stated that both causality and finality are 'necessary *regulative* (subjective) principles of thought and, as such, not mutually inconsistent' (Jung 1916/1917: par. 688). Phenomena must be considered from the point of view of both causality and finality:[8] 'To understand the psyche causally is to understand only one half of it' (Jung 1914: par. 398), since causality helps us understand the end result but fails to indicate the 'living meaning' (ibid.) of a process. To understand the whole of the psyche we have to view phenomena in the context of both approaches and simultaneously use methodologies that are suitable for each.

Jung's teleology 'does indeed go beyond causality, but it does not leave it behind' (Progoff 1973: 63). Even though teleology implies a final purpose, 'the means by which this implicit aim of life becomes actualized may quite well be described in terms of causal process; and from the other side, the conception of causality itself may quite acceptably be enlarged to include a teleological point of view' (ibid.). For example, we may explain a certain result (B) causally (i.e. A caused B) and concurrently – assuming we are open to the teleological modality – an underlying 'living meaning' could be perceived that actualized this process: B is not the result of A *per se*, but rather there is a contextual meaning – i.e. an overarching meaning – within which the event took place. Causality (i.e. efficient cause) and teleology (i.e. final cause) are two complementary points of view, both of which must be employed for a complete explanation of reality.

Jung was in the process of developing the theory that the psyche which functions predominately in accordance with the principle of teleology negates the rigid temporal parameters assumed by causality. Each principle implies a distinct notion of temporality. The temporally linear reality of consciousness, which presupposes a beginning and an end, is an unavoidable aspect of any process. However, the manner in which a beginning and an end are designated is not based on a concrete reality. Rather,

> We divide the processes from one another for the sake of discrimination and understanding, knowing full well that at bottom every division is arbitrary and conventional. This procedure in no way infringes the continuum of the world process, for 'beginning' and 'end' are primarily necessities of conscious cognition.
>
> (Jung 1934: par. 812)

Consciousness imposes a past that is conceived as the beginning of a process, and a future that is assumed to be the end-point. Yet, the reality of the psyche is not so clear-cut since 'beginnings and endings, merge into each other and form, strictly speaking, an indivisible continuum' (ibid.). In the context of everyday living a non-linear reality is translated into linear, with defined past, present and future for *the sake of conscious differentiation*. Once access is gained into the underlying world process this strict sequential arrangement evaporates and we can get a glimpse of a different temporal arrangement.

Jung assumed that there is constant interaction between past, present and future. An event in the past (which could be the source, for example, of a trauma or a memory) affects the present state. This, in turn, affects not only the future attitude but also the manner in which the past is understood. An understanding of consciousness based on linear modes of explanation is of limited value. The actual underlying world process is beyond linearity. In the same way that consciousness is assumed to be only a part of our total psychic activity (which is supplemented by unconsciousness), so a linear understanding of reality needs to be complemented by non-linear modes of understanding.

Papadopoulos, in an interpretation of Jung's epistemological standpoint, suggested that in order for the wholeness of reality to be understood we need to supplement the linear model with 'circular epistemology' (Papadopoulos 2006: 34). A linear model of causal reductive epistemology – where the past influences the present and the present influences the future in a sequential manner – is not in agreement with Jung's epistemology which favoured 'a constructive method that would highlight the function of "finality"' (ibid.). A circular epistemology is more appropriate since 'the pattern of mutual influence is the essence of Jung's alchemical model, where the circle (mandala) was the symbol *par excellence* of wholeness' (ibid.).

It seems however that a *non-linear* approach is more fitting in describing Jung's understanding than a circular epistemology. A circle is a line, albeit a never-ending one. The interaction between points in time is much more intricate and requires a rather paradoxical approach – at least according to the causally-driven conscious mind – whereupon temporality and causality are broadened. The supplementation of circularity with non-linearity and *multidimensionality* is also more aligned with the concept of the mandala, which, above and beyond its depiction as a circle, is a pattern.

Notes

1 Lightman 1993: 89.
2 'Freud's vocabulary, with its abundance of terms using a "re-/wieder-" prefix (reproduce, revise, remember, repeat, re-elaborate, etc.), is an unmistakable sign of his intense interest in the past. What we have to aim for is always a new version

of something that has already happened and did not go well; there is always something to "re-"' (Brizzi 1998: 60).

3 Even in the case *Nachträglichkeit*, which appears to make room for the future, the emphasis is on retrospection, a mechanism whose temporal directionality leaves no room for the future. The future dimension enters the theory in a roundabout way since it is not emphasized as such. The past is said to acquire meaning at a later point in time. The past is not re-created. The event stays intact; it is merely revisited and re-examined. See also Appendix.

4 'I have called this wholeness that transcends consciousness the "self". The goal of the individuation process is the synthesis of the self' (Jung 1940: par. 278).

5 See Appendix ('Freud on memory, the primal scene and the concept of *Nachträglichkeit*') for more details.

6 On several occasions, especially in his earlier writings Jung used teleology and finality interchangeably (see Jung 1916/1917: par. 688, 1916: par. 501*n*). However, it appears that he eventually created confusion around these terms by attempting to draw a distinction between them. It has been suggested that 'teleological aspects of the soul' found in Jung's earlier lectures were later replaced by the theory of 'finality of the unconscious and especially of the self' (Grivet-Shillito 1999: 96). Even though Jung conceived the concepts of finality and teleology as somewhat discrete, he nonetheless failed to clarify the qualities that would make such a distinction apparent. Most importantly, Jung clearly placed both finality and teleology at the opposite end of causality.

7 Jung clearly labelled the two kinds of cause – the final and efficient cause – while elaborating on Leibniz's theory which he utilized as support for the principle of synchronicity. Jung quoted Leibniz's explanation that 'bodies … "act according to the laws of *efficient causes*"… [whereas] souls act "according to the laws of final causes"' (Jung 1952c: par. 937). Following the citation Jung offered an explanation of Leibniz's theory by referring to efficient cause as 'causal connection' (Jung 1952c: par. 938).

8 'The psychic process, like any other life-process, is not just a causal sequence, but it is also a process with a teleological orientation' (Jung 1928b: par. 210). See also Jung 1916: par. 501*n*.

Chapter 3

Long-range effects in time

> Does it really exist, this destroyer, Time?
> When will it break the castle to shards
> on the calm mountain?
>
> Rainer Maria Rilke, *Sonnets to Orpheus*[1]

Expressions of the psyche's temporality: telepathy, dreams and foreknowledge

Jung's non-linear methodological approach found expression in phenomena such as telepathy, foreknowledge and dreams, as well as in the principle of synchronicity. In this chapter, as I attempt to delineate Jung's 'time theory', I take matters in a somewhat chronological order, starting with his writings on telepathy, which provide one of the earliest illustrations of his temporal understanding.

> The unconscious psyche appears to possess qualities which throw a most peculiar light on its relation to space and time. I am thinking of those spatial and temporal telepathic phenomena which, as we know, are much easier to ignore than to explain … The limitation of consciousness in space and time is such an overwhelming reality that every occasion when this fundamental truth is broken through must rank as an event of the highest theoretical significance, for it would prove that the space-time barrier can be annulled. The annulling factor would then be the psyche, since space-time would attach to it at most as a relative and conditioned quality. Under certain conditions it could even break through the barriers of space and time precisely because of a quality essential to it, that is, its relatively trans-spatial and trans-temporal nature. This possible transcendence of space-time, for which it seems to me there is a good deal

> of evidence, is of such incalculable import that it should spur the spirit of research to the greatest effort.
>
> (Jung 1934: par. 813)

Jung's early fascination with the concept of temporal and spatial relativity as it materializes via telepathic phenomena is exemplified with this somewhat lengthy quotation.[2] Indeed, telepathy seems to be the quintessential example of the relativity of space and time. The relativization of time and space demonstrated via telepathy leads to a re-examination of causality's role. As Jung put it, it is 'conditioned by space and time. But it demonstrates the relativity of space and time, and this is something I didn't invent' (1973: 327). Using telepathy as an observable example for the trans-spatial qualities of the psyche, he hypothesized that its temporal parameters should abide by the same laws: '[F]or philosophically considered, what do we mean by the "limitedness of space" if not a relativisation of the space category? Something similar might easily happen to the category of time (and to that of causality as well)' (ibid.: par. 814). The 'psychic limitlessness of time', as Jung unequivocally stated, is identical to its relativization.

Besides telepathy, the phenomenon of precognition was another example used by Jung to illustrate the soul's transcendence of time. In *The Zofingia Lectures* he suggested that instances such as premonitions, prophecies, second sights, clairvoyance and prophetic dreams are examples of 'long-range effects in time' (Jung 1897: par. 131). He, however, avoided elaborating on his observations: 'I will refrain from any attempt to explain long-range effects in time, despite the fact that the problem is vitally interesting and virtually begs for commentary' (ibid.).

Likewise, during his 1928–30 Dream Seminars although he appeared fascinated with children's future-telling dreams – 'Even most extraordinary problems can be dreamt by children, when one cannot see at all how a child could conceive of such things' (Jung 1984: 602) – he admitted unable to provide a theoretical understanding. He wondered: '[Do children] get [the knowledge of their future] through their parents? We still don't know. I think it must have to do with the collective unconscious, but that is another question which we will not go into now' (ibid.) Soon, however, Jung was able to investigate this phenomenon.

In the 1936–40 seminars on children's dreams Jung proposed that future-telling dreams are the product of the psyche's non-linear temporality. It was during this time that he articulated a hypothesis regarding the teleological, non-linear nature of time in the psyche. He returned to the problematic of causal interpretations of psychic processes and advocated for the replacement of 'strict causality by an interwoven action of conditions, to extend the unequivocal connection between cause and effect by a connection open to many interpretations' (Jung 2008: 4), reemphasizing the teleological nature of the psyche: 'We have to take into account that the psyche as such ... is of a goal-oriented, expedient nature' (ibid.).

The direction of the psyche is appropriate to an unconscious purpose, which is, nonetheless, unknown to the causality-dominated consciousness. While pondering the seeming lack of temporal dimension in dreams Jung spoke of the 'possibility' that:

> [T]here is something going on in the realm of the unconscious with the *notion of time*, that time comes apart a little in the unconscious, that is, that the unconscious always remains beside the passing of time, and perceives things that do not yet exist. In the unconscious everything is already there from the beginning ... The unconscious does not care about our time or the causal interrelation of things.
>
> (Jung 2008: 9–10)

Captivatingly, Jung characterized the above insight as quixotic. According to this 'quixotic possibility', the reason we apply a temporal order to dreams is because they enter consciousness in a sequential manner ('one after the other') and thus:

> [we] conceive them with the help of the temporal category and relate them to each other in a causal way ... The seemingly chronological series is, as it is, not the true series. If we conceive of it this way, we make a concession to our concept of time.
>
> (Jung 2008: 10)

In dreams, Jung concluded, conscious time is negated. Justifiably one might wonder why Jung was so reluctant in presenting what Freud had already established, namely the timelessness of the unconscious and, as an extension, of the dream.[3] In order to grasp the essence of Jung's position we need to adopt a broader perspective. He was not referring exclusively to the unconscious' temporality, but rather to temporality in the psyche *at large* – indeed, to 'our concept of time.' The false assumption of an existing linearly-temporal order in dreams – and with it, of causality – compromises *the true nature of time.* By adopting this approach we begin to appreciate why Jung appeared cautious when presenting his audience with a 'quixotic' claim; he was aware of the implications of such declaration. If time in the psyche as it had been understood was to be negotiated, he would have to find an alternative premise that would elucidate the temporal mechanism.

During the 1936–40 seminars Jung was still developing his understanding. Nonetheless, he provided an explanation that was under construction during this phase of his career. Dreams are not arranged in a linear chronological manner but rather in a *radial* manner. '[D]reams radiate from a centre, and are only later subjected to the influence of our time. In the final analysis, they are arranged around a *center of meaning*' (ibid.: 10). In effect, Jung was proposing a non-linear temporal modality of radial nature that would replace the causal interpretation (i.e. events causing each other).

A 'radially' arranged order implies a common centre. Jung advocated for the replacement of the causality-driven approach (which is based on linearity) with *a meaningful core* out of which the dream emanates. Causality would be expanded, or even replaced by meaning. References to this position can be found in his earlier work. While expanding on the constructive standpoint he suggested that the aim of this viewpoint 'is to elicit from the unconscious product a meaning that relates to the subject's future attitude' (Jung 1921: par. 702). The unconscious provides indicators of meaning that can explain our potential approach to life.

Such temporal modality implies *future meaning-making*; meaning is to be made: 'out there stimuli from the present interact with data gathered in the past, to allow prediction of possible futures' (Mathers 2001: 143). Meaning cannot be limited to the present moment, but is intimately connected to the psyche's teleological purpose, and thus – according to a linear understanding of temporality – to the future.

While exploring dreams that predict future psychological aspects unknown to the dreamer, Jung emphasized children's dreams that fall under this category perhaps to offer a dramatic example. It appears illogical that such dreams can anticipate the child's developmental process into adulthood since such futuristic images of a complete, as it were, state cannot pre-exist. Nonetheless:

> Strangely enough ... unconsciously the child already has all the psychology of an adult ... The individual is what it will be ... Such early dreams come out of the totality of the personality, and that is why they allow us to see a great deal of what we later miss in it. Later, life forces us to make one-sided differentiations.
>
> (Jung 2008: 20)

The temporal dimension of Jung's developmental understanding begins to emerge: our entire life-story, the whole personality, is present at the beginning. The future already exists in the present and is waiting to unfold; it has a *relative* meaning in regards to a linear view of time. Instances of precognition are most definitely not based on causality.

In order to adequately explain precognition Jung assumed that the psyche operates under non-linear temporality. 'When you are whole, you have discovered yourself once again, and you know what you have been all the time' (ibid.). That is, in the future (i.e. during adulthood) you will rediscover your wholeness, which was present since 'the past' (i.e. during childhood). If we interpret such foreknowledge from a linear temporal perspective this would imply that children are somehow capable of visiting a future point of their life-line, to engage in a sort of time-travelling. Causality, in this case, would be reversed. However, 'It seems utterly impossible to reverse causality, as in precognition, where a future event, apparently not yet in existence, causes an event in the present' (1976: 421) If we construe that the past, the present and the future are *not sequentially arranged* in the psyche, it is indeed possible for foretelling dreams to take place during childhood.

When we accept the possibility that the psyche is dominated by non-linear temporality, foreknowledge can be understood from a different perspective. Any future event is currently available in the present. This viewpoint presupposes a particular kind of time that is utterly different from the fixed/linear time, upon which causality operates. Specifically, if we consider the possibility that in the psyche linear time's demarcations all coexist in a relative state, then everything – and at the same time nothing – is indeed *simultaneous* and *contemporaneous*.

This understanding is in accordance with Jung's non-linear epistemology. A temporally-linear stand is replaced; any future meaning-making interacts and affects not only the present but also the past. Consequently, a principle that is based on teleological understanding is the only possible explanation of foreknowledge. Jung associated foreknowledge with nature's teleological mechanism: 'Final causes, twist them how we will, postulate a *foreknowledge of some kind*' (Jung 1952c: par. 931) He defined this foreknowledge as 'absolute knowledge' and explained that it 'is certainly not a knowledge that could be connected with the ego, and hence not a conscious knowledge' (ibid.).

The accessing of absolute knowledge overrides logical routes, i.e. the perceptual/sensory gathering of information that results in experiential knowledge. Absolute knowledge is the attainment of information that lies in the collective unconscious. Although the knowledge of our existence beyond conscious perception is unconscious and outside personal experience, at times it passes into consciousness and transcends an individual's experience so that 'one can very frequently observe an archetypal situation of collective nature' (Jung 1958: par. 849). Since the collective structures are independent of intellectual thinking and logical analysis, they have an inherent feeling of timelessness characteristic of the collective strata of the psyche – a supposition that is connected to the archetype's timeless and eternal temporality. And so, at those moments, a feeling of timelessness surfaces to consciousness and access is gained to the totality of our existence that lies beyond personal limitations.

Absolute knowledge can be conceptualized as the conscious attainment of timeless archetypal reality. Past, present and future – the linear temporal parameters of consciousness – are negated in the realm of the collective unconscious and a sense of foreknowledge, a by-product of the temporally-relative reality of the psyche, can be observed in consciousness. Absolute knowledge is the awareness of a profound all-encompassing realm beyond conscious perception and cognition. Incidences of absolute knowledge point the way to a potential future where a more self-aware life could unfold. Such knowledge seen from the ego's perspective appears absolute, since it transcends time and space and has a quality foreign to the limits of individual consciousness.

Exploring the psychophysical reality: the psychoid archetype

Jung's conceptualization of absolute knowledge is in accordance with his overall approach to the psyche. He was a psychologist who located his theories within the wider frameworks of understanding and expanded his vision beyond his assigned area of study. Jung's methodology was in search of a universal principle, one that would bridge the gap between physical laws and inner psychological reality. Matter is an intrinsic aspect of the psyche; in addition to the psychological dimension there exists a physical one.

Jung as Basilides explored the connection between the inner and outer reality early in his writings; he was evidently aware of the interdependence between the microcosm and the macrocosm. In Basilides' poem a meeting of the inner/intrapsychic and outer/cosmic worlds takes place in the psyche of man. Perhaps Jung's early observations left him unsatisfied with a purely psychological explanation. He felt compelled to include in his theory the physical aspect of nature since he detected interconnections between the material and the psychological realms. At an early point of his career he did so in a poetic manner, and not in the form of a concrete theory.

Jung introduced the notion of the *psychoid archetype* in order to articulate the view that the archetype belongs both to the material and the spiritual realm. The archetype has a dual nature that exists beyond the psychic level and extends to the material world so that 'the position of the archetype would be located beyond the psychic sphere ... and with its psychoid nature forms the bridge to matter' (Jung 1947/1954: par. 420). The concept of the psychoid archetype rests on the conclusion that a purely scientific or purely psychological approach cannot adequately explain the nature of reality, nor provide a sufficient explanation regarding the interrelation between outer and inner worlds.

Even though, '[m]atter and spirit both appear in the psychic realm as distinctive qualities of conscious contents' they, nonetheless, intersect and 'confront one another on the psychic plane' (ibid.). Consciousness is responsible for the discrimination of matter and spirit. Our efforts must be focused on bringing together the '*psyche and physical continuum*' (Jung 1947/1954: par. 440). And although 'science' has the same goal as Jung's approach, i.e. 'the grasping of the whole' (Jung 1952c: par. 864), it is impossible to achieve it via science alone since its approach is restricted by experimental and statistical procedures that do not allow 'Nature to answer out of her fullness' (ibid.).

Jung's unification of physical nature and psyche into a whole goes beyond repeatable experiments and statistics; even more so, it goes beyond a teleological (as opposed to causality-driven) epistemology. Bright suggests that Jung 'contrasts, rather, the scientific with the analytic approach' (Bright 1997: 615). The analytic method is at odds with the typical experimental setting where 'nature is ... artificially restricted in the laboratory to the question posed, and to that question alone' (ibid.). Such attitude does not embrace the whole of nature; it is one-sided. Conversely, the implications of his theory point to an approach that aims at understanding the workings of the world as a whole.

The drawback, however, of Jung's 'non-scientific' methodology, as he was well aware, is that 'one does not know what has happened' (Jung 1952c: par. 865). In other words, there is no solid evidence that would provide a tangible proof beyond scientific doubt. He offered the solution of the two Chinese sages who 'basing themselves on the hypothesis of the unity of nature, sought to explain the simultaneous occurrence of a psychic state with a physical process as *an equivalence of meaning*' (ibid). The two seemingly opposing domains of the physical and the psychic have meaning as their point of comparison; meaning is presented as an operative concept in the workings of nature. This approach served as the basis for the psychoid archetype. Jung would later indicate that such method can be observed via the principle of synchronicity.

A distinction is in order at this point. On most occasions when Jung criticized science he implied the investigative method, which is predominated by the one-sidedness resulting from the 'overvaluation of the "scientifically" attested views' (Jung 1947/1954: par. 426). Because of this approach the unconscious part of the psyche is neglected and the focus remains on the physical spectrum of reality:

> As a result of the prevailing one-sidedness, and in spite of the terrifying optical demonstration of an unconscious that has become alienated from the conscious, there are still vast numbers of people who are the blind and helpless victims of these conflicts, and who apply their scientific scrupulosity only to external objects, never to their own psychic condition. Yet the psychic facts are as much in need of objective scrutiny and acknowledgement.
>
> (Jung 1947/1954: par. 426)

For Jung the scientific approach implied the causal, reductive and positivistic attitude. His method, on the other hand, tried to encompass the whole of our existence, the unconscious as well as the conscious, the physical as well as the psychic. He believed – and this is where he departed from the mainstream theorists of his time – that the psychic condition can be reached through objective scrutiny and likewise, the physical can be accessed via psychological indicators and processes.

Jung implicitly differentiated between science which is restricted to the causal paradigm and science which takes into account non-causal relations as well as the role of the observer. These observations were (and still are) incorporated into the quantum paradigm. When Jung criticized science he was excluding the field of quantum physics, which operates on concurring premises as his own.

Main accurately wrote that the psychoid factor 'helps to bring about a rapprochement between psychology and physics' (Main 2004: 38). This statement, however, does not clarify which realm – the psychological examined by psychologists, or the physical examined by physicists – leads into the exploration of the other. Jung did not begin with an *a priori* accepted principle about a relationship between the psyche and the physical world. Rather, clinical experience led him to move beyond the workings of the inner psyche to the outer world. His observations left him unconvinced with a purely psychological explanation. He did not simply espouse the existing philosophies but rather he turned to the physical plane to achieve a more complete understanding of reality. The psychoid archetype was the product of his observations regarding the interrelation and interactions between these two realms.

The outstanding question of the space-time concept

Jung's venture to bridge the psychological and physical realms happened at an important point for physical sciences. Around the same time that he was introducing his theory, the field of physics was dominated by groundbreaking theories. After the Second World War science broke free of its existing assumptions on which its prevailing theories were based. The focus began to shift from the realm of macrophysics into that of microphysics. Jung observed these developments with fascination and was in contact (either through direct communication, or via third parties) with important figures in the field such as Wolfgang Pauli, Pascual Jordan and Werner Heisenberg among others.

With the discovery of quantum mechanics in 1925 Newtonian determinism was abolished and with it the hitherto scientific foundations which rested on linear time. A new paradigm, one that would allow for a principle other than causality to explain the physical world, was needed. These flourishing theories seemed to have sparked and informed his assumptions regarding psychophysical connections, and their temporal parameters. Jung's unique approach to the psyche allowed, and even invited, these additional sources of information beyond psychology, which would assist in the exploration of the physical aspect of the psyche. While developing his theory Jung consulted modern physics to gain support for his newly conceptualized thoughts.

In particular, Jung identified the metaphorical parallels between his observations and those of quantum physicists. He viewed the microcosm (which is the realm of observation of microphysics) as corresponding with the realm of the unconscious: 'The microcosm ... coincides with the unconscious' (Jung 1976: 421). He did not assume to possess knowledge of the science of physics. Rather he approached physics in a similar manner to the way he investigated other fields of knowledge: by extracting various symbols and metaphors in order to illuminate his understanding of the workings of the psyche; he utilized quantum physics as a metaphor.

At the same time, even though Jung unequivocally relied heavily on symbolism drawn from other fields to support his observations, he did not restrict himself to the metaphorical realm when taking the quantum theory into account. His remarks indicate that he detected theoretical similarities in his observations and those of contemporary physicists, and that he took the findings of the field into serious consideration:

> Physics has demonstrated, as plainly as could be wished, that in the realm of atomic magnitudes an observer is postulated in objective reality, and that only on this condition is a satisfactory scheme of explanation possible. This means that a subjective element attaches to the physicist's world picture, and secondly that a connection necessarily exists between the psyche to be explained and the objective space-time continuum.
>
> (Jung 1947/1954: par. 440)

Arguably, Jung's theory regarding the nature of the psyche, which included the material as well as the psychic realms, necessitated the information provided by physics, a field that investigates the material end of the spectrum. The findings of his contemporary physicists regarding the nature of time were of particular relevance to his observations; this was 'a specific step' that Jung deemed necessary for his theory. Jung was intrigued by the similarities of time-related aspects between the psychological and the physical realms. The observations by physicists threw light on his theorizations about time in the psyche and its interaction with physical reality, vis-à-vis the psychoid archetype. *Relative temporality* was a key aspect for the link between the material and the psychic realm.

When reviewing Jung's theory pertaining – explicitly or implicitly – to issues of temporality we frequently come across the utilization of physics for the development and fortification of an argument as demonstrated in the following passage.

> The hypothetical possibility that the psyche touches on a form of existence outside space and time presents a scientific question-mark that merits serious consideration for a long time to come. The ideas and doubts of theoretical physicists in our own day should prompt a cautious mood in psychologists too.
>
> (Jung 1934: par. 814)

Jung devoted considerable energy to the study of physicists' 'scientific question-mark' that pointed to time's relativity.[4] The behaviour of time was used as a departure point and an illustration of conformity between Jung and his contemporary physicists. According to Jung, Albert Einstein was the physicist who 'first started me off thinking about a possible relativity of time as well as space and their psychic conditionality' (Jung 1976: 109). Jung maintained that those interactions with Einstein 'exerted a lasting influence on my own intellectual work' (ibid.). Nonetheless, Wolfgang Pauli, one of the pioneers of quantum mechanics, was undeniably the most influential physicist for Jung's theory.[5]

Jung found Pauli's observations of the physical realm to be of particular relevance to his own psychological theories. He frequently used Pauli's explanations in order to amplify his view. Pauli proposed that the fields of psychology and physics are complementary disciplines and deal with the same fundamental principles. Their correspondence points to the fact that both Pauli and Jung believed they had found in each other a like-minded theorist.

Time had a significant role in the Jung-Pauli communication. As Jung attested in a 1948 letter to the physicist Pascual Jordan:

> [In *On the Nature of the Psyche*] we are discussing with Pauli the unexpected relations between psychology and physics. Psychology as might be expected appears in the realm of physics in the field of theory-building. The outstanding question is a psychological critique of the space-time concept.
>
> (Jung 1973: 494)

It is very likely that modern physics' critique of space-time was particularly attractive while Jung was developing his thinking regarding non-linear/relative temporality in the psyche. Pauli provided the scientific language for Jung's observations regarding time's behaviour in the psyche and challenged the causal and fixed nature of time by arguing for its relative existence in the material realm.[6] Pauli attested to the apparent conformity of opinions and the possible theoretical synthesis that could emerge from the fields of psychology and physics in regards to temporality. The way Jung's theory was developing, Pauli wrote,

> is indicative of a close fusion of psychology with the scientific experience of the processes in the material physical world. It is probably a long journey, one we are only just setting out on, and it will especially entail, as a modifying factor, constant criticism of the space-time concept.
>
> (Meier 2001: 32)

Previously Jung had associated his observations pertaining to the psyche's temporality with telepathy and precognition. In the field of physics, and specifically in Pauli's work, he found a credible scientific source that could amplify his understanding and increase his credibility regarding time's behaviour in the material realm.[7] Thus, even though Jung had already commented upon the relativity of time in the psyche[8] it was henceforth a certainty: 'We know positively only of the fact that space and time are relative to the psyche' (Jung 1954: par. 1574).

Despite the assertion that time in the psyche is relative, Jung indicated that *the conscious mind* needs definite time and space in order to grasp reality.

> Place and time are the most general and necessary elements in any definition ... A definite location in place and time is part of a man's reality ... When a definite time is added to a definite place one is rapidly approaching reality.
>
> (Jung 1936: pars. 283–5)

Here Jung's use of 'reality' refers to reality-as-conceived-by-consciousness. This interpretation is fortified by his accounts regarding fixed space and time as essentially non-existing: 'In themselves, space and time consist of *nothing*. They are hypostatized concepts born of the discriminating activity of the conscious mind ... [Space and time] are created by the intellectual needs of the observer' (Jung 1952c: par. 840).

Once again we encounter the temporal reality of the psyche as presented in *Septem Sermones*. On the one hand in the Pleroma everything, including time, is transcendent and thus relative; causality, by default, cannot be applied since events are not temporarily fixed. On the other hand, in Creatura time is absolute, and thus abides by the laws of causality. As his theory developed, Jung suggested that time in the psyche is relative; however, consciousness cannot grasp this relativity and thus conceptualizes time as fixed in order to account for its experiences of the outside world. The psychoid's temporality can illuminate the above position by providing an insight into the different manifestations of time in the psyche.

Time and the transcendent nature of the psychoid: the introduction of synchronicity

Jung's psychoid archetype marks an important departure from Freud's theory. Their diverse viewpoints become evident via their respective treatment of time. For Freud, the psyche rests on temporal dichotomy. The unconscious is ruled by timelessness, whereas consciousness is dominated by linear time. Jung, on the other hand, was a proponent of a unified field; the psychoid archetype was an attempt to explore this reality. I propose that Jung implicitly regarded time as *a unified temporal field* that manifests in different ways depending on particular circumstances. Thus, even though Jung, like Freud, spoke of timelessness that dominates the unconscious (collective and personal), for Jung – and here lies a vital distinction between the two theories – timelessness is not restrained in the unconscious. The emergence of consciousness did not compromise the psyche's temporality; *time is relative at large*.

The psychoid, Jung suggested, is inaccessible in its pure form:

> The archetype as such is a psychoid factor that belongs, as it were, to the invisible, ultra-violet end of the psychic spectrum. It does not appear, in itself, to be capable of reaching consciousness.
>
> (Jung 1947/1954: par. 417)

The real nature of the psychoid archetype cannot be made fully conscious because it is transcendent, that is, not available to immediate experience. Jung proposed the existence of a 'psychic scale' (ibid.: par. 367) that has '"psychoid" processes at both ends' (ibid.). On the one end of the spectrum exist physical processes and on the other end psychic processes. Beyond the threshold of consciousness lie psychoid phenomena, which are either – depending on which end of the spectrum they reside – quasi-physical or quasi-psychic.[9] However, the 'actual' psychoid is neither fully physical (biological/ somatic), nor fully psychical (psychological).

Even though archetypes as such cannot be consciously experienced due to their psychoid nature, we have an understanding of them via their manifestations and the way these affect our everyday experience: 'It is through the archetype that we come closest to this early "irrepresentable," psychoid stage of conscious development; indeed the archetype itself gives us direct intimations of it' (Jung 1976: 495). We observe the progression of the psyche and, consequently, its temporal dimension: out of the psychoid stage, consciousness develops. The psychoid archetype is the connecting link between an earlier stage of development where consciousness was not yet present and a more recent acquisition of the psyche, namely consciousness.

Since the real psychoid nature of the archetype is transcendent and thus cannot be made conscious, 'we have to rely solely upon qualities, that is, upon perceptible phenomena' (Jung 1947/1954: par. 417). Consciousness, in other words, allows for the *representation* of the irrepresentable psychoid nature.

There is a specific function for fixed time in the psyche: consciousness needs to *de-relativize time* in order to conceptualize it. Indeed, (to reuse Jung's poignant words), it appears that fixed time does not exist in itself but is 'only "postulated" by the conscious mind' (Jung 1952c: par. 840). The inconceivable-in-its-wholeness underlying reality, however, remains as is: undifferentiated by fixed – temporal, spatial etc. – categories.

The introduction of synchronicity appears to have been necessary for the elaboration of the psychoid. In addition, by investigating the manner in which consciousness deals with time we can come closer to the underlying connection between the psychoid and synchronicity. Previously we saw how Jung illustrated the psychoid as early as the writing of *Septem Sermones*. It was also argued that the precursors of synchronicity can be detected in the poem: man is a doorway where the two worlds (the macrocosm and the microcosm) meet. He incorporates these formulations into the principle of synchronicity: during synchronistic occurrences the personal and the archetypal realms unite. Furthermore, I suggest that the theory of synchronicity is an *application* of the psycho-physical nature of the psychoid. Jung's assertions support this view:

> [W]e must ask ourselves whether the relation of soul and body can be considered from this angle, that is to say whether the co-ordination of psychic and physical processes in a living organism can be understood as synchronistic phenomenon rather than as a causal relation … The synchronicity principle possesses properties that may help to clear up the body-soul problem.
>
> (Jung 1952c: par. 948)

Here synchronicity is presented as the principle that can assist in conceptualizing the 'body-soul problem'. Consequently, the advantage of considering synchronicity as an added principle to causality 'is that it makes possible a view which includes the *psychoid factor* in our description and knowledge of nature' (Jung 1952c: par. 961). Since the nature of the psychoid cannot be made fully conscious, we can employ synchronicity.

> Synchronicity tells us something about the nature of what I call the psychoid factor, i.e. the unconscious archetype (not its conscious representation!). As the archetype has the tendency to gather suitable forms of expression round itself, its nature is best understood when one imitates and supports this tendency through amplification. The natural effect of an archetype and its amplification can be certainly understood as an analogy of the synchronistic effect, inasmuch as the latter shows the same tendency of arranging collateral and coincidental facts which represent suitable expressions of the underlying archetype.
>
> (Jung 1955: par. 1208)

Synchronicity, similar to the psychoid archetype, draws together related expressions in the form of meaningful coincidences. However, unlike the psychoid, which cannot be made fully conscious, *synchronicity is observable.*

Thus, it can demonstrate by analogy the nature of the psychoid. This argument becomes clearer when this thought is considered:

> I incline in fact to view that synchronicity in the narrow sense is only a particular instance of general acausal orderedness – that, namely, of the equivalence of psychic and physical processes where the observer is in the fortunate position of being able to recognize the *tertium comparationis*.
>
> (Jung 1952c: par. 965)

Here two principles that fall under the umbrella of 'acausal orderedness' are implied: (a) synchronicity in the 'narrow sense' and (b) a different, and by default 'broader' kind of acausal orderedness.

In the former type of acausal orderedness, one is actually conscious of the commonality between the outer and the inner/psychic experience. There is an *observer* of the synchronistic experience who recognizes the existence of an archetypal orderedness in the natural world as well as in inner psychic reality; the *tertium comparationis* between inner and outer is detected. A synchronistic experience provides evidence for such orderedness. The latter type of acausal orderedness is a general type of acausal orderedness (Card 1991a), which is found in 'the properties of natural numbers or the discontinuities of physics' (Jung 1952c: par. 965) and can therefore be observed via scientific methods.

Jung conceived the principle of synchronicity as an *observable manifestation* of a wider, all-encompassing yet concealed principle which operates in nature at large. Even though we cannot truly know the underlying orderedness, we can get a glimpse of it through instances of synchronicity during which we can consciously observe a 'narrow' example of unified wholeness. Synchronicity makes the psychoid nature discernible.

The operative question, however, still remains: how does the relationship between synchronicity and the psychoid become manifested in Jung's theory? The role of time is of prime importance; synchronicity illuminates the psychoid reality when conceptualized in the context of time: it is an observable example of the *temporal* behaviour of the psychoid.

Chronos in synchronicity

In order to sufficiently elaborate the above hypothesis we need to investigate what synchronicity's relation to time is. This question becomes even more pertinent when we keep in mind that for Jung the psychoid archetype is capable of relativizing space and time, and that this quality becomes apparent via the principle of synchronicity: 'From synchronistic phenomena we learn that a peculiar feature of the psychoid background is transgressivity in space and time' (Jung 1976: 259).

Synchronicity is arguably the most recognizable term of analytical psychology, and yet there has been virtually no methodical treatment of it. It is as if most writers are seduced, or perhaps inspired, by the implications of synchronicity, which point towards a non-statistical, non-rational, and for some, spiritual realm, and therefore leave behind its systematic and scholarly examination. Also, it is surprising that, even though the principle's name has 'time' – *chronos* (*χρόνος*) – embedded in it very few authors have dealt with the concept in their work on synchronicity; and those who have, did not make it their primary focus.

Robert Aziz, in an exceptional treatment of synchronicity, suggested that, 'The time factor is unquestionably one of the most problematic aspects of Jung's thinking on synchronicity' (Aziz 1990: 63). He observed that time is unexplored in synchronicity; he attributed this to Jung's inadequate approach to the issue: 'Jung never did come to terms with the shortcomings of his discussion of the time factor with regards to synchronistic phenomena' (ibid.: 149). Perhaps the avoidance of the concept of time on the part of post-Jungians can be explained by Jung's own insufficient elaboration of the temporal implications of synchronicity. As a result, those who have dealt in depth with the principle have either barely touched upon the temporal element, or have treated it as an important yet unexplored parameter. Time in synchronicity remains an implicit assumption that has never been investigated in detail.

Despite the lack of systematic exploration of synchronicity's relationship to time, an examination of temporality in Jung's opus refers one to synchronicity, most likely due to its inherent temporal dimension unequivocally present in its name. I believe that this should hold true vice versa: in order to be complete an investigation of synchronicity would have to include its relation to time. The temporal dimension of the term (syn-chronicity) could not have been accidental; the word was chosen *because* of its emphasis on time; Jung deemed the temporal element vital for this principle. Furthermore, by systematically exploring Jung's time-related writings in the context of his collaboration with his contemporary physicists, parallel to the unfolding of the theory of synchronicity, the psychoid's hypostasis will become clearer; the link is time.

Even though Jung did not directly link the concept of time to synchronicity, he systematically evoked it in relation to the principle. His references to time were sometimes overt and sometimes implied. He spoke of synchronicity in its relation to simultaneity, a term that alludes to temporality: events occurring, happening or existing at the same time. As he indicated in a 1958 letter, 'Everything, at least so far as my experience goes, indicates that undeniable correspondences do occasionally occur, which I call "synchronistic" just because of their temporal simultaneity' (Jung 1976: 421) The principle of synchronicity 'asserts that the terms of a meaningful coincidence are connected by *simultaneity* and *meaning*' (Jung 1952c: par. 916).

Jung also referred to synchronism, a term closely related to the idea of simultaneity. Particularly in what appears to be his initial formulation of the principle – in a 1928 seminar note – Jung used the word synchronism to imply synchronicity:[10] 'Synchronism is the prejudice of the East; causality is the modern prejudice of the West' (Jung 1984: 44–5). Gieser points out that this first formulation of the term was an attempt to distinguish between simultaneity which is the 'typically Chinese' perspective and 'thinking in terms of "before and after", i.e. of cause and effect' (Gieser 2005: 277), which is characteristic of the Western way of thinking. Indeed, during a 1929 seminar Jung proposed that the difficulty in understanding synchronistic phenomena is strictly Western: '[It] is not difficult for the Chinese to accept; it is only difficult to our discriminating mind' (Jung 1984: 417). The use of the terms synchronism and simultaneity can be seen as initial attempts to put forward a non-causal approach.

In his more developed formulations Jung clarified that, even though it implies time via simultaneity, it is not identical to the term synchronous:

> I would like to call attention to a possible misunderstanding which may be occasioned by the term "synchronicity." I chose this term because the simultaneous occurrence of two meaningfully but not causally connected events seems to me an essential criterion. I am therefore using the general concept of synchronicity in the special sense of a coincidence in time of two or more causally unrelated events which have the same or similar meaning, in contrast to "synchronism," which simply means the simultaneous occurrence of two events.
>
> (Jung 1952c: par. 849)

Synchronous, in other words, does not include the factor of *meaning*, which is fundamental to synchronicity. Nevertheless, it is clear from the above quotation that the element of time remains central. Synchronicity – which presupposes a meaningful connection – must also involve a *temporal coexistence* (i.e. simultaneity) of two or more causally unrelated events.

This clause in the principle's premise has been criticized. Aziz proposed that the emphasis on simultaneity is misleading, since on several instances Jung himself provided examples of synchronicity where the inner and outer events are distant in time. Also, '[r]elated to this is the emergence of the erroneous notion that the time element constitutes the central unifying link in the synchronistic experience' (Aziz 1990: 70). Main (2007), as well, suggested that the requirement of simultaneity in the definition of synchronicity is problematic since in Jung's examples it is not a strictly necessary feature. Nonetheless, Main noted a function to simultaneity: if the outer and inner events are simultaneous their meaningful correspondence becomes more notable, and thus the observer is more likely to label the event as synchronistic. Simultaneity, in other words, even though not a necessary element, adds more clarity to the situation; it makes the meaningful correspondence more obvious.

As accurate as these observations might be, they seem to be based on a very strict definition of simultaneity. They operate from a linear time system, for which simultaneity implies co-occurrence of events. Justifiably, then, many post-Jungian theorists feel that 'the formulation of synchronicity in terms of simultaneity is ... riddled with difficulties' (Cambray 2009: 14). However, these criticisms seem to disregard the fact that Jung made an important clarification. There are *two sorts of simultaneity* involved in synchronistic phenomena: one is 'immediately obvious' (Jung 1952c: par. 850), whereas the other one is not. The distinction between these two types of simultaneity is due to the fact that, 'The unconscious ... often knows more than the conscious' (ibid.). Thus, events that would be characterized as 'obviously simultaneous' are the ones where consciousness has caught up, as it were, with the unconscious knowledge; for these synchronistic occurrences there exists, what Aziz accurately labelled 'clock-time simultaneity' (Aziz 1990: 71), i.e. the sort of co-occurrence that happens in conscious, linear time.

Conversely, events that are not perceived as simultaneous are the ones where the unconscious information is *ahead of consciousness* – so that one is not yet conscious of the simultaneity. In this latter category, even though simultaneity has not yet become known to consciousness, it nonetheless exists, albeit between the outer event and the unconscious – rather than between the outer event and consciousness, which is the case in the first category.

Given the apparent perplexity that exists around the issue of simultaneity, and the lack of attentiveness to the differentiation between its different categories, as well as each category's relation to time, it comes as no surprise that its role in synchronicity has been questioned.

Considering the above confusion regarding simultaneity, it would perhaps be useful to supplement the term with another word Jung used to describe the principle of synchronicity, which has not attracted much attention: parallelism. He indicated that, 'The strange cases of parallelism in time which are commonly called coincidences but which I call synchronistic phenomena, are very frequent in the observation of the unconscious' (Jung 1973: 177–8). Elsewhere he asserted that synchronistic phenomena 'consist in a parallel arrangement of facts in time' (Jung 1976: 46). He also referred to synchronicity as 'the simultaneous occurrence of a certain psychic state with one or more external events which appear as meaningful parallels to the momentary subjective state – and in certain cases, vice versa' (Jung 1952c: par. 850).

Parallelism, like simultaneity, falls into the category of implicit references to time. However, it has less of the connotation of 'at the exact clock-time' inherent in simultaneity. Moreover, simultaneity, as far as synchronicity is concerned, has no significance without meaning; it is not that two events are co-occurring, but rather that those events are connected by meaning. That was, after all, Jung's point when he spoke of the difference between synchronism and synchronicity. Parallelism has meaning embedded in it by definition since it refers to likeness, correspondence, or similarity in aspect, course or tendency.[11] In the context of synchronicity, it indicates that two events (in most instances outer and inner) are comparable or corresponding due to likeness, significant similarities or even exact agreement.

Evidently, even though Jung's time-related formulations vary, time as a factor remained a key element. As he came in contact with the way time was explored by modern physics he moved from implicit references to time (e.g. simultaneity and parallelism), to a more direct statement of its importance: he spoke straightforwardly of the relativity of time as a key parameter. Nonetheless, the role of simultaneity was not dismissed but, rather, it was refined. The fact that events are separated in time does not mean that they are separated in *relative time*; these presumed events could in fact be conjoined in the broader temporality of the psyche. Consciousness might not yet have registered this relation, but they are nonetheless connected in the unconscious – in the psychoid unconscious.

As the years passed, the relativization of time dominated Jung's formulations of synchronicity. As we have seen at the beginning of the chapter, Jung's initial conceptualization of temporality was greatly influenced by his interest in telepathy, which he had considered in a different context. The theory of synchronicity enabled him to recontextualize telepathy by linking it to the relativity of time, thus providing a more satisfactory explanation.

> [T]he phenomena of synchronicity ... are associated with the activity of unconscious operations and have hitherto been regarded, or repudiated, as "telepathy" ... The very diverse and confusing aspects of these phenomena are, so far as I can see at present, completely explicable on the assumption of a psychically relative space-time continuum.
>
> (Jung 1947/1954: par. 440)

The confusion that arises around the phenomenon of telepathy can be understood by taking into account the psyche's temporal (and spatial) relativity. Even though external reality might appear disconnected (i.e. two people might be located in different parts of the world) their psyches are connected in a temporally and spatially relative way. Assuming that time's behaviour relies solely on consciousness – which is 'isolated in its subjectivity' (ibid.) – makes phenomena such as telepathy appear inexplicable. If, however, we accept the possibility that there is a deeper underlying temporality that is relative then telepathic phenomena are 'completely explicable' (ibid.). Relative temporality provides the answer to otherwise incomprehensible phenomena.

Furthermore, time's role in synchronicity can offer an informative exegesis of the relationship between the nature of the psychoid and the principle of synchronicity. It may well be that Jung introduced the temporal element as a way of clarifying the distinction between 'general orderedness' of the psychoid and the particular form of orderedness apparent in synchronicity. Jung thought that there are eternal as well as systematic occurrences of orderedness that pre-exist in the world at large. General orderedness has 'existed from eternity and occur[s] regularly' (Jung 1952c: par. 965); in contrast, synchronistic occurrences are not eternal but are rather '*acts of creation in time*' (ibid.). By 'in time' Jung was implying the consciously perceived passage of time (i.e. clock time, linear time), which stands in opposition to eternity.

Jung offered a revealing clarification: 'That, incidentally, is precisely why I have stressed the element of time as being characteristic of these phenomena and called them *synchronistic*' (Jung 1952c: par. 965). For those pondering over the temporal emphasis of synchronicity this hardly qualifies as an incidental explanation. Manifestly, the word was purposefully chosen to highlight the role of *chronos*. Synchronicity indicates the all-encompassing eternal order due to its temporal quality. Since it occurs in time (i.e. it is 'an act of creation in time') it can be conceived and observed by our time-bound existence. Whereas eternity eludes us, synchronistic occurrences can be captured within 'our time'. Thus, even though Jung asserted the existence of the temporally relative nature of the psychoid, he believed that the materialization of synchronicity is *contingent upon time* – that is, fixed time. In fact, this is revealed by the etymology of the word he chose to describe this phenomenon: syn (*συν*) means not only with – which would simply imply simultaneity – but also 'with the help of' and 'in agreement with'.[12] Arguably, then, synchronicity unifies psychic and physical events, in agreement with fixed/linear time.

Observing meaning

In Jung's opus there are important time-related conceptualizations whose temporal dimension is, nonetheless, concealed. These aspects of his theory, examined in the context of the principle of synchronicity, can provide valuable information regarding the role of time in the psyche.

Jung described synchronicity as 'the simultaneous occurrence of a certain psychic state with one or more external events which appear as meaningful parallels to the momentary subjective state – and in certain cases, vice versa' (Jung 1952c: par. 850). Even though the observable events that are meaningfully related might not happen simultaneously in clock-time, they are experienced as happening in a particular 'momentary subjective state' (ibid.). This can be read to mean a certain phase in one's life, hence the characterization of it being subjective (i.e. belonging to/taking place in one's mind, rather than being objective). At such times, something – i.e. a theme, a pattern, an issue, a constellated archetype – is active so that one notices certain types of experiences which correspond with (or are parallel to) certain outer events. In my view, this is precisely what Jung meant when, in his 1929 formulations of the concept of synchronicity he stated: 'I have invented the word *synchronicity* as a term to cover these phenomena, that is, things happening at the same moment as an expression of the same time content' (Jung 1984: 417). Outer and inner events are connected *in time* since they belong to 'the same time content' of one's life.

This postulation is related to the factor of meaning: events belong to 'the same time content' because they have the same meaning. Jung emphasized that 'meaning is an anthropomorphic interpretation … What that factor which appears to us as "meaning" may be in itself we have no possibility of knowing' (Jung 1952c: par. 916). It is our *perception* that identifies events that are acausally connected as synchronistic rather than the fact that they occur at the same time; meaning is given to these events by an observer. (This is the same formulation expressed in 1916 by Jung as Basilides: we project our inner differentiated reality and in turn we observe this differentiation as part of the external reality.)

What is perhaps less apparent is that this approach does imply a causal relationship of sorts. Even though there is no causal link *per se* between the events, there is nonetheless an *inner cause* that connects them, namely the meaning attributed to them by the observer. In this sense, *meaning is, in and of itself, a cause*. The cause is one's own meaning-attribution. Events are connected because meaning is given to them by the one who observes the situation. However, the making of meaning in a synchronistic instance is *not* causality-dependent in the ordinary sense, since it is not based on linear time.

Meaning seems to be the *sine qua non* of the principle of synchronicity: 'The idea of synchronicity' has an 'inherent quality of meaning' (Jung 1952c: par. 961). Only those chance coincidences that are meaningful are considered synchronistic. Based on this premise Aziz theorized that meaning is of prime importance for synchronicity whereas Jung's 'time factor' is of questionable value: time 'is not the unifying factor of the psychophysical archetypal field; the unifying factor is simply meaning' (Aziz 1990: 149).

However, there are additional dimensions that are equally important as meaning, which are, nonetheless, so intricately connected that they seem to appear as one factor. One of these is the role of the observer who attributes meaning to the non-causally connected events. The other vital aspect is consciousness. Meaning cannot be made unless the observer of the psychophysical archetypal field is conscious: 'Without consciousness there would, practically speaking, be no world, for the world exists for us only in so far as it is consciously reflected by a psyche' (Jung 1957: par. 528). Moreover, there is the parameter of time. All of the above-mentioned factors (the observer, consciousness and time) are essential for synchronicity's making of meaning. The unifying factor in the theory of synchronicity is thus not 'simply meaning', but rather *meaning derived in conscious time by an observer.*

Taking the well-known example of the golden scarab,[13] the presence of the beetle at the window was not meaningful by itself, neither was the dream of the golden scarab. These were mere occurrences (inner and outer). What granted them meaning was the patient's conscious observation at a particular point in time. As Morgan states, although the keyword of synchronicity is meaning, 'it is the meaning that consciousness finds in these occurrences that give significance. It is what "fixes" the event' (Morgan 2000: 124–5). The observer, who perceives the inner and outer events, attributes significance to the occurrence and, thus, deems it synchronistic.

Jung's theory regarding meaning-making and the role of the observer is full of the quantum paradigm. As Pauli explained in a 1947 letter, modern physics challenged the hitherto Newtonian conviction that space and time were 'God's right hand' (Meier 2001: 32) and instead placed power in the observer's freedom of choice. Evidently, Jung was aware of the parallels between his hypothesis pertaining to the observer's crucial role in synchronicity and in quantum mechanics; he utilized the findings of his contemporary physicists to add credibility to his theory of synchronicity.

> [T]he latest conclusions of science are coming nearer and nearer to a unitary idea of being, characterized by space and time on the one hand and by causality and synchronicity on the other … [This finding] seems

> to show that there is some possibility of getting rid of the incommensurability between the observed and the observer.
>
> (Jung 1952c: par. 960)

The relationship between the observer and the observed as investigated by physics was of particular assistance to Jung's development of the principle of synchronicity. His interest in these findings becomes evident in his comments on the quantum physicist Pascual Jordan's writings on telepathy. Jordan was interested in the relativity of space in parapsychological phenomena, and particularly in 'spatial clairvoyance'. Pauli sent Jordan's work to Jung, since he considered it to be of relevance to Jung's theory.[14] Jung, after being introduced to Jordan's writings on 'spatial telepathy', commented on the results. He agreed with Jordan that our empirical image of space is nonexistent, and that spatial clairvoyance is an obvious representation of that. He then drew a parallel between the role of the observer in quantum mechanics and his own observations:

> Jordan's approach would lead to the supposition of an absolute unconscious space in which an infinite number of observers are looking at the same object. The psychological version would be: In the unconscious there is just one observer, who looks at an infinite number of objects.
>
> (Meier 2001: 7)

Jung suggested to Jordan (via Pauli) the inclusion of temporality in his investigations: 'Taking [Jordan's argument regarding spatial relativity] further, he would also necessarily have to bring in temporal clairvoyance, which would represent the relativity of the image of time' (ibid.).

In order to better understand Jordan's position, and in turn Jung's suggestion, it would be useful to take a closer look into the quantum paradigm. Werner Karl Heisenberg (1901–76), one of the founders of quantum mechanics, composed the theory of Quantum Indeterminism: for a particular quantum state there are infinite possible futures and realities. Once, however, an observer obtains a measurement only one result is detected; upon observation a single observable reality is materialized out of an infinite range of possibilities. The one observable reality is not caused by the observer, but rather it is the one reality that materializes out of numerous possibilities.[15] Gammon (1973) noted the parallel that exists between Heisenberg's uncertainty principle and synchronicity. Internal or external events in space-time that are assumed to be causally connected are not so; they appear to be connected, but they simply existed as possibilities that are charged by the observer's attention. The observation 'prompts' the events to acquire meaning, which can be mistaken as a causal connection between the events – since the event and the observed outcome happened around the same time, or one following the other.

The function of the observer in the principle of synchronicity cannot be overstated. Also, on a metaphorical level, we can detect the similarity between 'the meaning-giving observer' in the principle of synchronicity and 'the substance-giving observer' in the quantum field: quantum physics suggests that material reality depends on the observer and thus it is subjective; the mere observation of nature alters the item under scrutiny by materializing only one probable outcome. In Wolfgang Pauli's words, microphysics is based on the principle that, 'The observer [is] a little lord of creation in his microcosm, with the ability (at least partially) of freedom of choice and fundamentally uncontrollable effects on that which is observed' (Meier 2001: 32). The observer, upon consciously examining the 'object' or (in the case of the principle of synchronicity, the inner and outer events), 'causes' one probability to materialize *in fixed space and time.* Events are located in a particular space-time *because of* conscious observation.

Correspondence on a complex and vexing idea

The parallels between the field of quantum physics and Jung's approach could provide an illuminating exegesis for time's role in synchronicity. Thus it would be useful to study in some detail Pauli's contribution to Jung's ideas. Pauli's influence on Jung's time theory reached beyond the recognition that time is relative. Pauli was one of the intellectual stimulators for the theory of synchronicity. In a letter dated 22 June 1949 Jung wrote to Pauli:

> Quite a while ago, you encouraged me to write down my thoughts on synchronicity. I have finally managed to get around to it and more or less collect my thoughts on the subject. I would be most grateful if you would be kind enough to cast a critical eye over it, covered as it is with question marks. Nowadays, physicists are the only people who are paying serious attention to such ideas.
>
> (Meier 2001: 36)

With this letter started a lengthy correspondence between the two men regarding the principle of synchronicity. Hogenson described their communication as 'unique in Jung's corpus; it may be the only instance where Jung is genuinely attentive to, indeed deferential to, a correspondent over an extended period of time while dealing with the most complex and vexing of his ideas' (Carter and West 2008: 127). Jung's principle of synchronicity was formed through this interchange; Pauli's opinions served as a measure of the theory's effectiveness and acceptability. Jung would later attest to the close connection between his 'relation with the physicist Professor W. Pauli and ... my thesis of psychic synchronicity' (Jung 1976: 109).

Beyond his overall remarks on synchronicity, it appears that Pauli was 'largely responsible' (Meier 2001: lix) for Jung's 'strange discovery' (Jung 1973: 494) of the *quaternio* around which the principle of synchronicity was elaborated. In Jung's *quaternio* synchronicity was the fourth principle, in addition to the existing 'triad of classical physics' (Jung 1952c: par. 961) consisting of space, time and causality. Jung drew this parallel:

> Just as the introduction of time as the fourth dimension in modern physics postulates an irrepresentable space-time continuum, so the idea of synchronicity with its inherent quality of meaning produces a picture of the world so irrepresentable as to be completely baffling.
>
> (Jung 1952c: par. 961)

For Jung the added dimension of time by physics was equivalent to the addition of synchronicity. The emphasis on time was in no way accidental.

Pauli's initial resistance to Jung's formulations of the *quaternio*, as well as Jung's response, can further illuminate the role of time in Jung's theory. It seems that Pauli was in disagreement with Jung's placement of time in opposition to space in the *quaternio*. In a 1950 letter he explained that a 'modern physicist would find [the placement of time opposite space] particularly unacceptable' (Meier 2001: 56). Pauli argued that '[time and space] do not form a true pair of opposites (since space and time can easily be applied simultaneously to the phenomena)' (ibid.) There is a space-time continuum and, in Pauli's view, it should remain intact.

Pauli therefore surmised a 'quaternary scheme' of physics in which space and time were not placed in opposition. He proposed that in physics 'indestructible energy and momentum' are in opposition to a 'definite spatial-temporal process'. Similarly, in psychology there exists an equivalent pair, namely the 'timeless collective unconscious' which stands in opposition to 'self-awareness, time' (Meier 2001: 57). For Pauli, each of the psyche's realms is dominated by a different sort of time. He offered the following *quaternio* (ibid.):

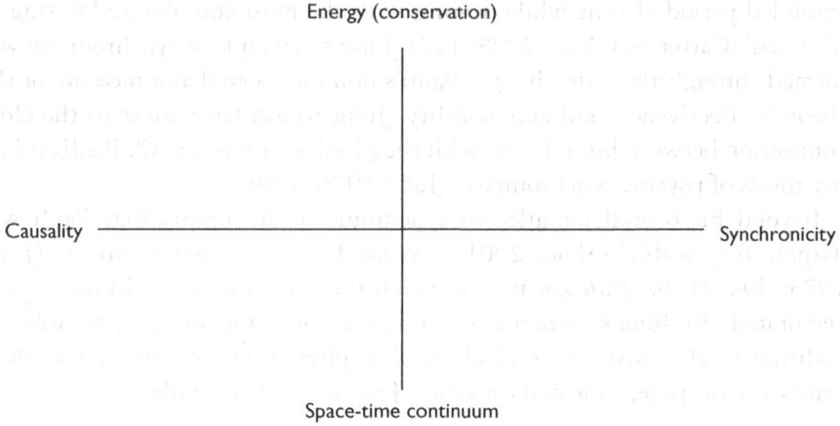

Jung agreed and indicated that the space-time continuum is applicable to psychology as well:

> Insofar as psychology penetrates into the unconscious, it probably has no alternative but to acknowledge the 'indistinctness' or the impossibility of distinguishing between time and space, as well as their psychic relativity.
> (Meier 2001: 60–1)

Jung thus accepted that even though space and time appear to be 'eternally separate and antithetical' (ibid.), they are in fact united in the psyche. This formulation brings in mind the Gnostic understanding of pairs of opposites: syzygies, such as Space and Time, do not stand in opposition but rather complement and complete each other.

In Jung's maturing theory causality and synchronicity 'on an empirical and practical level (i.e. in the real world) … are incommensurable and antithetical, like space and time' even though they are 'identical in the higher sense of the term "connection" or attachment' (Meier 2001: 60–1).

It is important to observe once more that Jung did not contain timelessness in the unconscious realm. He spoke of the 'empirical and practical level' (ibid.) that stands in opposition to the 'higher level'. Jung's position becomes clearer in light of the revised *quaternio* (Jung 1952c: par. 963):

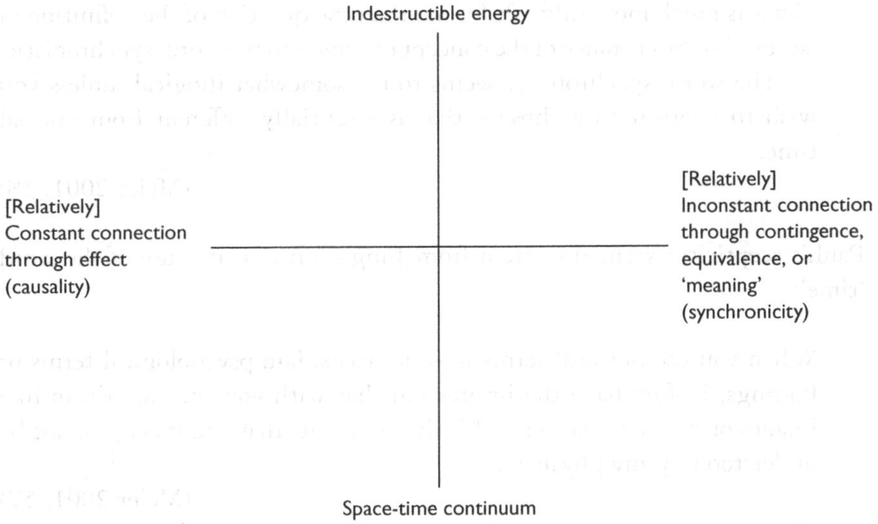

Events are connected by causality *and* synchronicity within the 'space-time continuum' *and* within the realm of permanence (i.e. 'Indestructible energy'). In the case of causality (or perhaps more accurately of 'efficient cause') events are connected by 'effect': they are brought about by a cause; they are the result/consequence of something else. In the case of synchronicity, events are connected by meaning. Furthermore, meaning, being subjective (i.e. dependent on one's personal experience) is relatively 'inconstant', whereas efficient cause, being objective, provides a relatively constant connection. It seems that meaning *is* 'the higher sense of the term "connection" or attachment' (Meier 2001: 60–1). Thus, connection is relatively inconstant – perhaps this is why the word connection is in quotation marks in the above quote. Cause-and-effect, on the other hand, being dependent on 'an empirical and practical level (i.e. in the real world)' (ibid.) could provide a 'relatively constant connection'.

Pauli's remarks regarding the inseparability of the space-time continuum were incorporated in Jung's final manuscript on synchronicity. Jung regarded this *quaternio* as 'perfectly suitable for a more complete world-picture' (Jung 1952c: par. 963) and maintained that, 'This schema satisfies on the one hand the postulates of modern physics and on the other hand those of psychology' (ibid.: par. 964).

However, this claim appears to be an exaggeration since Pauli was at odds with the way Jung used the term 'time' in his theory. In Pauli's words:

> What is much more difficult for me than the question of the definition of 'acausal' is the entrance of the concept of time into the word 'synchronistic' ... The word 'synchron' ... seems to me somewhat illogical, unless you wish to relate it to a chronos that is essentially different from normal time.
>
> (Meier 2001: 38)

Pauli's scepticism seemed to stem from Jung's 'unscientific' use of the word 'time':

> When you use physical terms in order to explain psychological terms or findings, I often have the impression that with you they are dreamlike images of the imagination ... [Such statements that you make] cannot be understood by any physicist.
>
> (Meier 2001: 57)

Jung, in his eagerness to find a psychological principle that corresponded to the recent and fascinating findings of physicists, utilized concepts in an incorrect way, at least according to physics' definitions.[16]

Perhaps Pauli's apparent dissatisfaction with Jung's utilization of the concept of time could be avoided had time been properly discussed and defined. Perhaps this could be achieved by emphasizing the existence of two utterly different uses of the word time in their correspondence (and in Jung's opus at large): fixed time that dominates consciousness, and relative time that exists in the psyche at large. Furthermore, fixed time could be identified with the 'space-time continuum' of the final version of the *quaternio* (Jung 1952c: par. 963) whereas relative time would be identified with the realm of 'Indestructible energy' (ibid.). Also, relative time could be linked with 'dreamlike images' (Meier 2001: 38) proposed by Pauli, since it would take into account the imagistic/ symbolic aspect of the psyche. This, in turn, is time on the 'higher level', which stands in opposition to time on the 'empirical and practical level' (ibid.: 60–1). This opposition exists literally in the *quaternio*: the two kinds of time are found on opposite ends of the cross.

Perhaps, had Jung clarified – what Pauli had suspected – that he was in fact talking about 'a chronos that is essentially different from normal time' (ibid.), and had he provided an adequate definition of the temporality on which the reality of the psyche operates, such misunderstanding could have been avoided. Nonetheless, although Jung did not make the distinction clear, in his explanations of synchronistic phenomena he did mention time in a particular way, i.e. 'a *chronos* that is essentially different from normal time':

> It looks as if the collective character of the archetypes would manifest itself also in meaningful coincidences, i.e., as if the archetype (or the collective unconscious) were not only inside the individual, but also outside, viz. in one's environment, as if sender and percipient were in the same psychic space, or in the same time.
>
> (Jung 1976: 46)

Jung implied a temporal (as well as spatial) parameter that behaves *as if in conscious time*, but is nonetheless different from it. Synchronistic occurrences, even though they seem to happen due to or in accordance with our time-and-space-bound reality, occur due to factors that are not contingent upon consciously-perceived time or space. Had Jung clearly defined this distinction such confusion could have been avoided.

Jung was aware of the fact that physicists' time was fundamentally different from time for psychologists. As he wrote in 1953, while commenting on his relationship with Einstein, 'One can scarcely imagine a greater contrast than that between the mathematical and the psychological mentality. The one is extremely quantitative and the other just as extremely qualitative' (Jung 1976: 109). The following year he explained that while 'causality is a *statistical truth*' events of an 'acausal nature bordering on the category of synchronistic ... have to do with "qualitative time."' (ibid.: 176).

Gieser offered an illustrative comparison between the Western and Chinese conceptualization of time, which can further illuminate the kind of distinction Jung was making. For the Western mind there exists an 'arrow of time', where cause precedes effect. In opposition to this stands a 'field of time':

> Instead of an arrow, a field is more a suitable analogy here. The centre of the field is a moment in time around which certain events are grouped ... as if time behaved as a magnet which attracted certain events. But the events are not causally associated with each other and time does not cause their grouping; rather the events express the *quality* of the moment in time. Time is not merely an abstract parameter, a before and after, which links cause and effect. Time is above all a *quality*, expressing a pattern which has a significance or meaning.
>
> (Gieser 2005: 279)

As we have previously seen, Jung suggested that instead of a linear arrangement (i.e. an arrow of time) there exists *a centre of meaning* around which dreams are arranged (Jung 2008); meaning could be conceptualized as the common quality that groups dream-events together. Events in the dream are not arranged chronologically, but rather meaning connects them qualitatively.

This can also apply to Jung's time theory: Events (inner and outer) occur in a non-causal manner in the context of a temporal field that gathers together events of similar quality, that is, of similar meaning. Jung's analytical approach was already based on the idea of clusters of meaning, an idea that goes all the way back to the Word Association Experiment and is certainly apparent in the idea of complexes. So it is not surprising that this way of thinking contributed to the conceptualization of the psychoid archetype and the principle of synchronicity; the idea of a centre of meaning extends into the realm of time, in which relative temporality is also conceived as a 'field of time'.

Notes

1 Rilke 1960: 109. Reprinted by permission of the University of California Press.

2 In the years that followed Jung wrote on telepathy: 'Although parapsychological problems lie on the extreme edge of my field of research, I still follow with great attention the new developments which one frequently encounters in psychology in regard to these phenomena' (Jung 1973: 494).

3 Freud observed that dreams compress conscious time: 'A dream is able to compress into a very short space of time an amount of perceptual matter far greater than the amount of ideational matter that can be dealt with by our waking mind' (Freud 1900: 64). During the 1911 meeting of the Vienna Psycho-Analytic Society Freud remarked that via the dream we can observe the unconscious' 'misorientation with regard to time ... that is, present, past, and future are one', as well as the fact that in the unconscious there is 'absence of the effects of the passage of time' (Freud, as cited in Nunberg and Federn 1974: 308).

4 Other writers have indicated that the theories of Jung and those of many twentieth-century physicists complement each other: see Nichols 1980; Peat 1988; Card 1991b; von Franz 1992; Zabriskie 1995, 2001.

5 See for example Lindorff 1995; Meier 2001; Gieser 2005; Carter and West 2008; Miller 2009.

6 General Relativity as expressed by Einstein's formulations holds a different position on time than quantum mechanics, which was Pauli's main field of interest; they operate under different temporal assumptions.

7 Jung was evidently aware that scientific language could facilitate in conveying his theory: 'So long as you keep to the physical side of the world, you can say pretty well anything that is more or less provable without incurring the prejudice of being unscientific, but if you touch the psychological problem the little man, who also goes in for science, gets mad' (Jung 1973: 177).

8 See for example Jung 1936: par. 175; 1939, par. 684; 1948, pars. 747 and 754, Jung 1951a: par. 254. He would repeat this statement on several occasions (e.g. Jung 1957: par. 527).

9 'I use [the term psychoid] to distinguish a category of events from merely vitalistic phenomena on the one hand and from specifically psychic processes on the other' (Jung 1947/1954: par. 368). The term psychoid was used by Jung to describe processes that are between 'somatic life-energy and sheer bodily process on the one hand and true psychic processes on the other' (Stein 2003: 96).

10 McGuire noted that 'Apparently [this is] Jung's first use of this term in the sense of "synchronicity," or meaningful coincidence, as an explanatory principle of parallel physical and psychic events, equal in importance and complementary to the principle of causality' (Jung 1984: 44-45*n*).

11 *The American Heritage Dictionary of the English Language*, 2003.

12 Bambiniotis 1998

13 According to Jung (1952c: par. 982), while a patient was narrating her dream of a golden scarab an actual scarab tapped on his office window. Jung argued that if the possibilities of causality and of a mere coincidence are excluded then a different force must be behind this event; he suggested that the occurrence was an example of the workings of the principle of synchronicity.

14 Pauli characterized Jordan as a gifted but troubled physicist (Meier 2001). This is a curious portrayal of a man who happened to be not a mere 'gifted but troubled' scientist, but one of the physicists who set forth the theory of quantum mechanics in 1925 (along with Max Born and Werner Heisenberg), a theory that would revolutionize science – curious, especially if one considers the fact that such a description came from Pauli.

15 There is no consensus among quantum physicists as to how this one reality takes place, but many theorize that it is related to the consciousness of the observer: the observation itself promotes that single actuality (Davies 2002).

16 A student of Pauli later stated that synchronicity 'is something which physicists do not know about, nor would they wish to' (Peat 1988: 14). Peat commented: 'His implication was clear: synchronicity smelled of pseudo science and loose thinking' (ibid.).

Chapter 4

A unified reality

All life desires eternity.

C. G. Jung, *Memories, Dreams, Reflections*[1]

Meaning is unequivocally emphasized in the principle of synchronicity. So far we have considered meaning's relationship to the element of time via the role of the observer. In order for meaning to be derived consciousness must be employed; a conscious observer must attribute meaning to acausally related events. It is necessary to clarify this: meaning is not invented by the observer. The meaning observed in synchronistic phenomena 'presupposes . … a latent meaning which can be recognized by consciousness' (Jung 1976: 495). During synchronistic occurrences the observer is capable of consciously detecting the *underlying meaning* embedded in the psychoid reality.

Yet, the psychoid is transcendent and thus cannot be made fully conscious. The all-encompassing and transcendent 'Meaning' (with a capital M) becomes detectible through 'meaning' (with a lower-case m) bestowed by the observer. The pre-existing latent Meaning of the psychoid remains hidden in both the physical and the psychic aspects of nature and yet manifests itself via synchronistic phenomena. Thus meaning, even though made actual because of consciousness, is restrained by its limitations.

The other requirement besides meaning is 'a psychoid process with which a physical event meaningfully coincides' (ibid.). However – and this is where the primary importance of consciousness, and with it of meaning derived 'in conscious time' comes into play – 'Here [i.e. during a psychoid process with which a physical event meaningfully coincides] the meaning cannot be recognized because there is as yet no consciousness' (ibid.). The recognition of pre-existing latent Meaning assumes consciousness, and it is only within a time-bound state that consciousness can observe it. The observer is 'at the right place, at the right time' and thus is able to detect the otherwise latent Meaning that is inherent in the psychoid reality.

In Jung's theory there are two distinct kinds of time: relativized time of the psyche, and fixed time of consciousness. The two realms are not austerely separated in their temporalities. During a synchronistic event, contents move out of relative space-time and into the space-and-time-bound world of consciousness. With this movement the archetype carries into the realm of consciousness a sense of timelessness and eternity. This experience provides a feeling of absolute knowledge. Absolute knowledge is the knowledge of psychic correspondence with the outside world, or as Jung put it 'of the presence in the microcosm of macrocosmic events' (Jung 1952c: par. 923). Synchronistic experiences exemplify the existence of the psychophysical continuum's inherent Meaning. The continuum operates within the bounds of conscious time but – due to its acausality and its access to the deeper atemporal collective structures of the psyche where Meaning resides – brings forth a sense of knowledge beyond the temporally and spatially restricted world.

This is connected with Jung's teleological approach. Inherent Meaning lies dormant until it is manifested in the *telos*, i.e. at the end of the process (as for example in the case of children, where the meaning of their dreams is to be consciously revealed in adulthood). However, such interpretation could be misleading since it relies on linear time progression. Rather, latent Meaning is *constantly present*; it is all-encompassing. It is a matter of consciously tapping into that moment when, at a particular time and a particular space, ever-present Meaning becomes apparent to the diligent observer who, in turn, attributes meaning to the moment.

Perhaps, then, Pauli's metaphorical conjecture that 'The observer [is] a little lord of creation in his microcosm, with the ability (at least partially) of freedom of choice and fundamentally uncontrollable effects on that which is observed.' (Meier 2001: 32) needs to be re-examined. The observer does not create the cosmos (micro or macro); the observer is merely observing the macrocosmic level, on a microcosmic scale. At a particular moment – which Jung identified as synchronistic – the observer consciously experiences, albeit momentarily, the interconnection of nature and gains access to Knowledge of the Cosmos, i.e. to absolute knowledge. At that moment the psychoid reality, which reaches beyond the limits of consciousness and into a holistic and unitary reality, is observed.

Jung explained instances of foreknowledge as experiences of absolute knowledge since at those moments access is gained to a reality that exists beyond the personal realm. However, 'foreknowledge' – like 'teleology' – is an unfortunate characterization since a particular future point (fore), and thus linear temporality, is implied in the etymology. Instead, the concept of absolute knowledge is associated with ever-present Knowledge – perhaps identical to ever-present Meaning evoked in instances of synchronicity. Through absolute knowledge access is granted to undifferentiated reality – a reality that is beyond temporal and spatial demarcations.

The sense of undifferentiation that is associated with instances of absolute knowledge can exemplify this argument. During the golden scarab incident Jung's patient came into contact with a powerful image, which was inexplicably (at least for the conscious mind) connected with the presence of an actual scarab in Jung's office. Her dream came at a pivotal point: when she was in a position to observe these events in conscious time (during therapy). It provided her with a sense of the world beyond her conscious and intellectual control, and carried into consciousness an archetypal image (i.e. the golden scarab). She had a sense of absolute knowledge, which surpassed the limitations posed by consciousness; she became aware of the intimate connection between her personal experience and the outer world. Importantly, especially for this particular patient who Jung described as overly intellectual, she became aware of the unconscious' influences in her life.

The feeling of undifferentiation does not derive solely from synchronistic occurrences; it often accompanies the experience of being in love. During those instances we feel 'beyond' or 'besides' ourselves; boundaries are expanded in a transpersonal manner as we merge with the object of our affection. Eros transports us to a state beyond ego-boundaries. Emily Brontë in her novel *Wuthering Heights* beautifully described the sense of spilling of one's self into another.

> [B]ecause he's more myself than I am. Whatever our souls are made of, his and mine are the same ... I cannot express it; but surely you and everybody have a notion that there is or should be an existence of yours beyond you.
> (Brontë 1953: 106–8)

This description is reminiscent of Freud's oceanic feeling: '[A] a feeling which [is associated with] a sensation of "eternity", a feeling as of something limitless, unbounded – as it were, "oceanic" ... That is to say, it is a feeling of an indissoluble bond, of being one with the external world as a whole' (Freud 1930: 64–5). Even though he attached this description to religious experiences, it can aptly apply to the feeling of being in love.[2] The oceanic feeling allows for the possibility that the timelessness ascribed to the unconscious could be experienced while conscious.

Marie Bonaparte wrote of love as an 'outlet' from time, equal to dreaming, religion and intoxication. When in love 'even when awake [we are allowed] to breathe the enchanted atmosphere of the dream ... In its power to evoke this lies the greatness of love. Every lover, however wretched his condition may otherwise be, finds himself transported into fairy-land' (Bonaparte 1940: 434). Undifferentiation is attached to a 'sensation of "eternity"' (Freud 1930: 64). When in love we function from the unconscious where time is lost; we are taken back to 'the childhood illusion that for us time knows no bounds' (Bonaparte 1940: 434). This could be perceived as 'the glory of love' since when under its spells we live 'the supreme moments in which time, and the need to grow old, are really forgotten. That is why every lover swears eternal love' (ibid.).

Bonaparte allowed for the transcendence of our time-bound existence via love – a possibility that Freud rejected, albeit indirectly. In *The Interpretation of Dreams* Freud wrote that Frau B.'s dream of meeting the man she was in love with the night before the event took place could not be considered prophetic; the dream was produced by a suppressed thought. Freud concluded: 'Accidents which seem preconcerted like this are to be found in every love story' (Freud 1900: 624). Had he allowed for the possibility that every love story is characterized by such 'preconcertions' precisely *because* being in love alters something in our temporal experience, he could have accepted Frau B.'s description as authentic.

Bonaparte spoke of the feeling of love as a magical solution to the bonds of time. Jung as well gave love immense power (albeit in a less romantic tone) by referring to 'love-magic' as the sort of enchantment that 'exercises a *compulsion* that prevails over the conscious mind and will of the victim: an alien will rises up in the bewitched and proves stronger than his ego' (Jung 1951b: par. 216). This kind of magic is comparable to 'the compelling power' (ibid.) of unconscious contents.

When in love we function as if in a dream: not from the ego, but rather from an unconscious state, where time has different connotations and its linear flow is abolished. It is indeed an enchanting state, since even though awake we can experience the 'magic' of unconscious temporality. In the case of Catherine, Brontë's heroine, through her love for Heathcliff she accessed a psychic space beyond conscious individual boundaries. Through the archetypal experience of love she bypassed logic and gained access to the transpersonal realm where consciousness is experienced in a diffused way. She reached beyond her individual life and death – 'If all else perished, and *he* remained, I should still continue to be' (Brontë 1953: 108) – and, consequently, beyond the material world. She experienced the union of opposites beyond inside-outside, life-death, me-you; she arrived at absolute totality beyond conscious time.

Undifferentiation lies at the foundation of absolute knowledge. Jung had adhered to a unified reality of the psyche since the early stages of his career when in *The Zofingia Lectures* he spoke of the principle of *All in One*; an account reminiscent of the Gnostic undifferentiated Pleromatic state. Jung associated this world order with the concept of the *World Soul* and the *unus mundus*. Even though he attached different meanings to each of these terms there is a fundamental premise that unites them: the existence of an overarching unified energy that 'shapes all things' (Jung 1952c: par. 931).

Jung explained that this holistic approach was dominant for centuries until 'natural science proved man's subordination to nature and his extreme dependence on causes' (Jung 1952c: par. 928). Western philosophy during the last three centuries[3] with its emphasis on isolation of the mind has divorced man from his 'primordial oneness with the universe. Man himself has ceased to be the microcosm and eidolon of the cosmos' (Jung 1939/1954: par. 759). Since intellect and conscious perception alone cannot comprehend such profound all-encompassing reality, man has banished such ancient worldviews in favour of causality. However, despite the causality-driven worldview of rational consciousness, absolute reality is accessible. We contain within us the blueprint of the whole and can thus access it by employing faculties beyond intellectual thinking and causal explanations. Jung's approach offers a possible solution to humanity's estrangement from primordial oneness.

For Jung the separation from pre-existing oneness is not simply the result of Westernization. Rather, the distinction into different realms is necessary for our limited perception. Humans require the artificial separation into psychological and physical in order to cognitively grasp reality; the existence of pairs of opposites as described by Basilides indicates the same principle. Consciousness necessitates the duality of psyche and matter, as well as the separation between observer and object. It also demands defined (as opposed to relative) space and time: 'Cognition ... presupposes an opposition, a here and there, an above and below, a before and after' (Jung 1961: 308). This duality is also a prerequisite of synchronicity: 'synchronicity in order to be observed requires two essentially heterogeneous world-systems, whose sporadic interlocking causes certain aspects of wholeness to manifest themselves' (von Franz 1974: 247).

Observations of 'unitary dualism' (ibid.) can only be made in conscious time. Yet through observation, contemplation and meaning-making of synchronistic events in conscious time the psychological and the physical meet. As Jung indicated, '"speculation", a very typical form of consciousness, becomes the real centre of the world, the basis of the *Unus Mundus*' (Jung 1976: 470). In consciousness the observation of a unified reality takes place;

the physical and the psychological words unite in the contemplative qualities of consciousness. Paradoxically, the recognition of the unified reality of the *unus mundus* requires consciousness' duality.

The psychoid archetype, Jung reminded his reader,

> points to the sphere of the *Unus Mundus* ... Although the first step in the cognitive process is to discriminate and divide, at the second step it will unite what has been divided, and an explanation will be satisfactory only when it achieves a synthesis.
>
> (Jung 1958: par. 852)

There is much to suggest that Jung's own attempts to 'unite what has been divided' culminated with the principle of synchronicity. He explicitly associated synchronicity's meaningful connections with the concept of the *unus mundus*: '[The] principle [of synchronicity] suggests that there is an interconnection or unity of causally unrelated events, and thus postulates a unitary aspect of being which can very well be described as the *unus mundus*' (Jung 1955–6: par. 662).

Underlying primordial oneness implies that the meaning derived from synchronistic occurrences is not an artificial product of the human intellect. Rather, 'those things in which this spirit [of the World Soul] is particularly powerful ... have a tendency to "beget their like" in other words, to produce correspondences or meaningful coincidences' (Jung 1952c: par. 931). The unified reality itself, which contains the whole of nature, produces meaningful coincidences which are in turn perceived by an attentive observer. With synchronicity Jung reclaimed humanity's role as a microcosmic element of the *unus mundus* and place the psyche in the cosmic realm where it once belonged.

Symbolic elaborations

Even though time is an integral element of Jung's theory he did not include it in his reflections on the *Unus Mundus*. However, von Franz elaborated on *unus mundus*' relation to time by utilizing the World Clock, a symbol that consists of one vertical and one horizontal circle with a common empty centre.

The World Clock holds an odd place in Jung's writings. According to Jung, this symbol was dreamed by one of his patients. Curiously, even though he amplified the dream on different occasions, he did not mention that it was one of Pauli's visions shared with Jung during their private communication. Rather, Jung described it in the vague context of 'a patient's dream'. It is through the correspondence of the two men (as opposed to Jung's official writings as found in the *Collected Works*) that we come across a comment by Pauli regarding this particular dream.[4]

Jung elaborated on specific details of the image – such as its different colours and rhythms (Jung 1938/1940: par. 111) and offered various amplifications of the symbol. He paid particular attention to its centre that is empty as opposed to other mandalas (e.g. the horoscope and the actual clock). Jung concluded that it must be 'a small scale model or perhaps even a source of space-time, or at any rate an embodiment of it and therefore, mathematically speaking, four-dimensional in nature although only visible in a three-dimensional projection' (Jung 1936: pars. 311–12[5]). However, he offered no further comments. This would have been a rather suitable point to at least mention the fact that the World Clock was a symbol described by Pauli, the physicist with whom Jung had detailed exchanges regarding space-time.

Jung considered an elaboration on the supposition that the World Clock might be a symbol of the 'source of space-time' to 'lie beyond [his] powers of proof' (Jung 1936: par. 312). A possible explanation for his reluctance to further comment on this hypothesis might be the fact that Jung's conversations with Pauli regarding the space-time continuum and the *quaternio* happened at a later date. Perhaps at this time he was unable to rely on his own 'powers of proof'. Jung did not return to this symbol in his later writings.

Von Franz – who, captivatingly, maintained that Jung's efforts to further explore his thoughts were hindered by his humility – used the World Clock symbol to suggest that it is a 'double mandala structure ... which ... in fact represents a kind of primordial symbol of the Unus Mundus with space-time generated out of it' (von Franz 1992: 56). During an earlier account she had suggested that the motif of the double mandala is both timeless and temporal. The common centre of the two circles – in the case of the World Clock, of the horizontal and vertical ones – is 'the moment of contact [which occurs] when a spontaneous action emanates from their common centre' (von Franz 1974: 262). The double mandala encompasses the wholeness represented by the *unus mundus*.

> [At those moments when] eternity breaks through momentarily into our temporal system, the primal unity actively manifests itself and temporarily unites the double structures into one, so to speak. This is how the *unus mundus* becomes revealed in the phenomenon of synchronicity. But immediately afterward the flow of events resumes its course on the track of the ordinary temporal pattern, and the timeless order falls back into a latent condition once more.
>
> (ibid.: 263)

The intersection of the two circles symbolizes the observation of a synchronistic experience; at 'the moment of contact' (ibid.: 262) the timeless and the time-bound intersect, psyche and matter meet and their artificial split no longer exists. When a synchronistic experience is spontaneously observed, the time-bound human gains access to the eternal and timeless realm, which lays latent until the moment of contact. During synchronistic experience a sense of wholeness overcomes ego-bound cognition and absolute knowledge is attained.

However, the portal to eternity and this numinous experience does not last long; just long enough for the observation to be made and for its psychological impact to be recorded by consciousness. Then time gains its accustomed sway and we return to the bounds of linear, fixed and causal time. To use Basilides' language, the portal into the macrocosm is shut once more and man is restricted in his microcosm, bound by the limitations of Creatura.

Since the *unus mundus* represents the unified reality of the psychoid world it can be viewed as the container of conscious and unconscious parts of the psyche alike and, consequently, of their corresponding temporal realities. Even though psyche and matter are of two separate natures they 'are contained in one and the same world, and moreover ... [they] are two different aspects of one and the same thing' (Jung 1947/1954: par. 418). Once consciousness-based temporally-bound knowledge has been attained we must return to the whole, which is the only true expression of Reality. The *unus mundus* is the point of departure, as well as the end point of all; it enfolds the psyche's beginning as well as its end, its *telos*. At the same time it is eternal and beyond time.

Similar to the World Clock, the opposition contained in the *unus mundus* appears to find its symbolic expression in the Uroboros, the tail-eating serpent, which depicts an all-embracing reality. Jung referred to the Uroboros on numerous occasions primarily in connection with alchemy. Even though he did not directly connect the symbol with his understanding of temporality, it may help in appreciating his 'time theory'.

The Uroboros represents opposing forces such as 'the deadly and life-giving water' (Jung 1942/1954: par. 359) and opposing creative powers such as 'active and passive, masculine and feminine' (Jung 1938/1954: par. 105). It encompasses the unification of opposing elements and is thus 'a symbol uniting all opposites' (Jung 1937: par. 404). Jung, in accordance with the Gnostic tradition, suggested that opposing forces are complementary to each other: 'Every psychic phenomenon is compensated by its opposites, in agreement with the proverb, "Les extrêmes se touchent"' (Jung 1933/1934: par. 292). The touching of the serpent's extremities – its tail bitten by its own mouth – seems to be a fitting representation of this premise: the manner in which the head relates to its tail 'describes the opposites which are secretly one' (von Franz 1980: 116).

The serpent is 'a chthonic and at the same time the "most spiritual" animal' (Jung 1951b: par. 389). In the symbol of the Uroboros the opposition of the 'lowest, most material' and the 'highest and most spiritual' (ibid.) unite. The psychophysical continuum becomes an endless and beginningless ring. To indicate the unification of all opposites Jung identified the Uroboros with 'the Greek alchemical formula of εν τό πάν' (Jung 1942/1954: par. 353), i.e. the All One, the one that contains the whole/everything. Through this symbol the psychoid's original wholeness is restored.

While exploring the development of Jung's time theory two contradictory temporal modalities were revealed: linear time which is related to consciousness, and non-linear temporality associated with the psyche as a whole. However, these modes of time are based on completely different premises. How can time be both linear and non-linear? Linear time implies a beginning and an end-point and, therefore, a sequential order of events. These restrictions do not apply to non-linear time where events coexist; there is no definite order, no beginning or end. I have argued that Jung attempted to link these two opposing modes of time via the principle of synchronicity. Nonetheless, he did not sufficiently muse on the concept of time even though it appears to be of vital importance. Perhaps his attempts halted because he could not find a theoretically concrete way to connect the two paradoxical natures of time.

As Jung noted, once faced with a paradox, a symbol can convey the contradiction much better than a logical argument.[6] The Uroboros can illuminate this temporal inconsistency. Jung stated that 'The uroboros is one even though in the twilight of the unconscious its head and tail appear as separate figures and are regarded as such' (Jung 1955–56: par. 402). The symbol and the archetype of primordial oneness that it represents are depicted as phenomena that dwell at the borderline between consciousness and the unconscious. Even though in consciousness the extremities of the serpent are distinct, in the unconscious they regain their archetypal unity. Applying the contradictory modes of temporality to this image, in the realm of consciousness time is linear and thus the serpent appears as a horizontally placed animal with a concrete beginning (its head) and a distinct end (its tail). However, in the realm of the unconscious the Uroboros regains its eternal force as 'a magic circle consisting of the union of opposites, "immune to all injury"' (Jung 1951b: par. 418). In the unconscious it becomes 'the eloquent symbol of the Infinite Eternal One' (De Rola 1997: 14).

Perhaps the Uroboros' dwelling in the 'twilight of the unconscious' (Jung 1955–56: par. 402) refers to the quasi-conscious state of the contents of the unconscious, which seem to possess a kind of consciousness.[7] In a rare mention of time in relation to the Uroboros Jung suggested that:

> The all-seeing dragon is a symbol of time, related to Chronos. The Uroboros has the meaning of eternity (αἰών) ... The identification of the All-Seeing with Time probably explains the eyes on the wheels in Ezekiel's vision ... We mention this identification because of its special importance: it indicates the relation between the *mundus archetypus* of the unconscious and the 'phenomenon' of Time – in other words, it points to the *synchronicity* of archetypal events.
>
> (Jung 1947/1954: par. 394)

In Jung's opus the motif of the all-seeing – *polyophthalmia* (many eyes) – is associated with 'multiple consciousness' (Jung 1950: par. 690), that is, with the various quasi-conscious states existing in the unconscious. Multiple eyes 'seem to be a typical symbol of what I have called the "multiple luminosities of the unconscious"' (Jung 1955–56: par. 170). Particularly, *polyophthalmia* 'indicates the observing consciousness' (Jung 1950: par. 690) of the unconscious. This interpretation of the all-seeing motif concurs that consciousness is the observing agent of the psyche.

Jung identified the Uroboros with eternity, which is the temporal modality of the archetypal world and the unconscious. Eternal and thus non-linear time is intimately related to the all-seeing 'phenomenon of Time'. Taking this conceptualization further, I would argue that 'all-seeing Time' refers to the *omnipresence of non-linear temporality*; hence, Time is capitalized.

If something is all-seeing it makes possible the simultaneous observation of various perspectives, in the absence of a particular sequence. If various perspectives are/can be perceived at the same time then, by default, linear perception is excluded. Thus, the 'all-seeing Time' symbolized by the Uroboros signifies non-linear multiplicity of perspectives.

Perceived in this light, the Uroboros does not depict cyclical time but rather the lack of linear time. This, in turn, provides additional support for the previously presented hypothesis that cyclicality, even though a significant departure from expressing time linearly, it still relies on linearity and unidimensionality. This symbolic expression emphasizes the reality of primordial oneness: in a unified reality the past, present and future coexist and can be observed *simultaneously*; 'multiple eyes would indicate a multiplicity of conscious centres which are co-ordinated into a unity' (Jung 1955–56: par. 271).

Polyophthalmia can symbolically illuminate the concept of foreknowledge. It is not about knowing something in advance ('fore'), but rather about being able to observe what is already in existence via a simultaneous multiplicity of perspectives. In turn, the Uroboros provides a useful image for the paradoxical coexistence of linear and non-linear time in the psyche. Temporality as symbolically portrayed by the Uroboros remains unrestricted by its circular shape: it is simultaneously time-bound, non-linear and eternal.

Notes

1 Jung 1961: 304.

2 Freud explained the feeling of oneness as the ego's need to control the pressures of the external world. The id's sense of immortality and timelessness is connected to feelings of infantile omnipotence, which are disturbed by the ego's perception of time. For the child, the passage of time symbolizes the period of separation. This is contrasted with the feeling of timelessness that is interconnected with the fantasy of endless unification with the mother. We have a somatic experience of the oceanic feeling, and to this we long to return. (For more details see Bergler and Róheim 1946; Jacobson 1954; Namnum 1972; Gifford 1960, 1980; Masler 1973; Hartocollis 1974; Calhoun 1976; Morris 1983; Arlow 1986; Priel 1997; Rose 1997).

3 According to Jung's accounts, 'during the last two centuries' (Jung 1939/1954: par. 759). Even though a century has elapsed since Jung's writings this statement still holds.

4 Pauli (1938): 'In summer I came across your book *Psychology and Religion*, and I saw that you have attached a certain importance to some of my early dreams, especially the vision with the "World Clock"' (Meier 2001: 20). Meier added in a footnote: 'Pauli is not alarmed by anonymous use of his material, nor does he argue with the interpretation' (ibid.: 21*n*).

5 See also Jung 1938/1940: pars. 112–18.

6 'Any content that transcends consciousness, and for which the apperceptive apparatus does not exist, can call forth the same kind of paradoxical or antinominal symbolism. For a naïve consciousness that sees everything in terms of black and white, even the unavoidable dual aspect of "man and his shadow" can be transcendent in this sense and will consequently evoke paradoxical symbols' (Jung 1942/1948: par. 277).

7 'The archetypes have about them a certain effulgence or quasi-consciousness, and the numinosity entails luminosity' (Jung 1947/1954: par. 388).

Chapter 5

The death of time

Mythical consciousness and the descent to the archaic

> [Y]our ancestor did not think of time as absolute and uniform. He believed in an infinite series of times, in a dizzily growing, ever spreading network of diverging, converging and parallel times. This web of time – the strands of which approach one another, bifurcate, intersect or ignore each other through the centuries – embraces *every* possibility.
>
> Jorge Luis Borges, *Ficciones*[1]

Death in the Jungian paradigm: a symbolic perspective

In the beginning of our exploration of temporality in Jung's opus I proposed that his basic premises were present in his early writings. Even though he had not yet developed the theoretical implications and intricacies of the workings of time, the same temporal principles were present since the beginning, albeit in a less developed form. After exploring temporality in Jung's theory I will attempt to re-examine his initial formulations from a more informed perspective.

Jung began his time-related musings with symbolic elaborations. He represented time with the libido, and suggested that death and its opposing energy, life, symbolize time itself; the fundamental link between time and libido is their oppositional qualities. Time, being an elusive concept, can only be symbolized by an energy that captures its opposition.

> Time is ... defined by the rising and setting sun, by the death and renewal of libido, the dawning and extinction of consciousness ... [T]ime, this empty and purely formal concept, is expressed in the mysteries through transformations of the creative force, libido.
>
> (Jung 1911–12/1952: par. 425)

Having in mind the unfolding of Jung's 'time theory', I would consider an additional amplification of this symbolic presentation. His formulations are not limited to one kind of time. 'The dawning of consciousness', which brings with it the awareness of death, can be identified with the realm of linear time and, conversely, 'the extinction of consciousness' is a reference to the realm of non-linear time.

This, in turn, would imply that not only is time symbolized by the libido but also that time is *transformed through the opposition* of life and death. Different temporal expressions are not viewed as two separate, incompatible states, but as two opposing temporal forces that interact and produce the creative force. In these early accounts Jung, by providing the symbolic background for the 'empty and purely formal concept', initiated the exploration of the psyche's inherent temporal opposition.

Jung's contemplations on death and life have far-reaching implications regarding temporality.[2] Death and life are primarily symbolic representations of primordially opposed energies that transform the psyche's energy. Jung's libido which encompasses 'both the life and the death forces, leads to greater attention to the potentially creative nature of conflict. In consequence, not opposition but complementarity tends to be emphasized' (Gordon 1977: 111). Through creative opposition wholeness is achieved. 'The Mother of Death and the Mother of Life in lamenting the dying god, and, as an outward token of their union, Mary kisses the cross and is reconciled' (Jung 1911–12/1952: par. 415). On the cross – the quintessential symbol of wholeness[3] – the definitive opposites of life and death unite; opposition gives way to wholeness. Jung continued: '[T]hrough Adam's sin, which lay precisely in his becoming conscious, death came into the world' (ibid.) The temporal dimension of this amplification transpires: out of eternal recurrence of unconsciousness, opposition was born due to Adam's sin. Consciousness created linear time, which brought with it death.

Nonetheless, death is a desired condition.

> [The hero's] life is a constant struggle against extinction, a violent yet fleeting deliverance from ever-lurking night. This death is no external enemy, it is his own inner longing for the stillness and profound peace of all-knowing non-existence, for all-seeing sleep in the ocean of coming-to-be and passing away. Even in his highest strivings for harmony and balance ... he seeks death, immobility, satiety, rest.
>
> (Jung 1911–12/1952: par. 553)

In this passage stillness can be seen as the death of time, as the end of change and movement. Change and movement belong to linear temporality. This kind of time has, thus far, been associated with consciousness. By 'the working of time … everything young grows old, all beauty fades, all heat cools, all brightness dims, and every truth becomes stale and trite' (ibid.). Jung followed this thought with a paradoxical suggestion: '[All things] age, sicken, crumble to dust – *unless they change*' (ibid.: par. 553; italics added). Linear time, which is one and the same as change – the time whose passage crumbles one to dust – can be defeated *with change*.

> But change they can, for the invisible spark that generated them is potent enough for infinite generation. No one should deny the danger of the descent, but it *can* be risked. No one *need* risk it, but it is certain that some one will. And let those who go down the sunset way do so with open eyes, for it is a sacrifice which daunts even the gods. Yet every descent is followed by an ascent; the vanishing shapes are shaped anew, and a truth is valid in the end only if it suffers change and bears new witness in new images, in new tongues, like a new wine that is put into new bottles.
>
> (Jung 1911–12/1952: par. 553)

Jung favoured a deviation from the adherence to time's arrow. He offered an antidote to death that is the outcome of change: the descent rewards those who dare with infinite generation. And even though infinite implies a spatial quality, Jung's words point towards the realm of eternity, a realm *beyond time.*

Mythical and eschatological consciousness

If we are to understand the manner in which the descent could provide liberation from linear time, we must move away from an eschatological perspective. Eschatology, in so far as it is perceived as an idea or a doctrine concerning the ultimate or final state (of things, beings, eras etc.), does not leave room for a perspective beyond an imminent end. However, if we explore the concept symbolically a new understanding would transpire; a symbolic standpoint allows for a more tolerant interpretation, one that does not imply an inescapable conclusion. Such a standpoint could overhaul the perspective where the impending 'end of the world' (one's personal world or the world at large) looms.

The work of Helmuth Plessner could offer assistance. He spoke from an anthropological point of view; he wrote of historic time, of the development of the experience of time throughout history.

> A cyclically ordered world knows death only as an organic phenomenon … [I]t is only with the transformation of the mythical consciousness of time from its cyclical to its eschatological form that a vision opens on the 'nevermore' and a separation arises between past, present, and future.
>
> (Plessner 1957: 236)

Plessner placed cyclical time outside the unconscious realm, and into the realm of *mythical consciousnesses*. Once mythical consciousness has been transformed into eschatological consciousness death arises as a problem. This newer form of consciousness, which is concerned with the idea of the end (of the world), created separation between past, present and future; eschatological consciousness created linear time. We can discern two kinds of consciousness operating under opposing temporalities: eschatological consciousness, which is based on linear time, and mythical consciousness, which is based on cyclical time. The latter kind bears resemblance to the realm of the unconscious. Perhaps then Jung's proposed descent does not necessitate plunging into unconsciousness. It would be sufficient to enter the realm of mythical consciousness.

Plessner continued:

> Primordial time, the immemorial 'once upon a time' which lays down the prototypes for all happenings, does not narrow down into a 'time of beginning' until there is a corresponding 'time of the end'. This *transition* from a world without beginning and end to a world of extreme limitations is clearly connected with a release of man from 'nature.'
>
> (Plessner 1957: 237)

Speaking from a historical perspective, Plessner was referring to the removal of humankind from primitive ways and into modern life. A literal removal of ancient cultures from nature – i.e. from the natural world that enclosed primitive people – was responsible for this shift. He considered this 'release' as a positive step in humanity's development, and linked it to our stepping onto 'a new plane'.

Bearing in mind the temporal implications of each kind of consciousness it transpires that 'man's release from nature' was responsible for the transition to a world where death awaits at the end of the line. Thus, if we were to reclaim the connection to nature we could avoid the threat of eschatology. Also, the return to nature could be conceptualized as a return to what lies within rather than a return to the natural world (our habitat, the trees and caves). We can thus conceptualize Jung's descent as follows: We must revisit our *inner nature*;

we must reconnect to the part of our psyche that has been abandoned for the sake of an eschatological perspective, which promised to raise us to a new plane and instead has plunged us into a death-fearing existence.

The descent to the archaic

This approach is related to Jung's *Archaic Man* (1931). The archaic is not only a chronological quality but it describes 'a kind of thinking' (Bishop 2008: 504), characterized by a state of being where there is no distinction between inner and outer worlds. Archaic temporality is a fusion of linear time of 'outside' reality and cyclical temporality of our 'inner' experience. It is constantly accessible – and thus can be accessed by the modern man because 'the "archaic" is not simply very, very old – it is essentially timeless' (ibid.). Bishop's use of 'timeless' in the above extract seems to be qualitatively the same as Freud's timeless unconscious: it is unaffected by the passage of time rather than being outside it; the archaic quality remained unaltered *through time*. At the same time the archaic way of being implies a temporal quality that is *outside time*; it is truly time-less, indeed eternal: '[T]he quality of eternity which is so characteristic of the life of primitive humankind, is entirely lacking in our lives. Isolated and hemmed in by our rationalistic walls, we are cut off from the eternity of nature' (Jung 1928/1931: par. 739).

Returning to the amplification of Plessner's account, this is the primal nature from which man has been 'released'. However, in light of the Jungian paradigm, a separation from our primordial state seems more of a burden than liberation. There is a heavy price to pay for the elevation to 'a new plane'.

This, in turn, is linked to the psychoid's psycho-physical wholeness. Jung offers a way towards such reconnection through analytical psychology, which paves the way for the return to the timeless archaic within. Through the above amplifications we can grasp the meaning of Jung's calling for the descent, which offers liberation from the 'constant struggle against extinction' (Jung 1911–12/1952: par. 553). In view of these elaborations, 'the descent' seems like an apt metaphor, i.e. the descent to a lower plane, where inner nature is still connected with outer reality.

This interpretation involves a temporal dimension that is in accordance with Jung's 'time theory'. The rediscovery and return to the archaic parts of our psyche is also 'a way forward to the future: we need, in other words, to (re-) engage with what one might call *le futur archaique*' (Bishop 2008: 518). In a footnote accompanying the oxymoron of *le futur archaique* (the archaic future), Bishop writes: 'This intriguing expression nicely captures the full chronological ambiguity of Jung's sense of the archaic: at once historical, timeless, and more full of presence than the present' (ibid.: 518*n*). With this note Bishop illuminated the temporality of the archaic: it is simultaneously historically ancient – in regards to our phylogenetic history – and also eternally present in the unconscious parts of the psyche. The notion of the archaic with its timeless quality as well as its historical significance can lead us into the future. The future is the linear progression of the present historical moment, and simultaneously, according to Jung's time theory, it is always present since it is eternal and timeless.

Similar to the call for a reconnection to the archaic, James Hillman advocated for a return to the 'true nature' of the child; that is, to *the imaginal realm*. Since the imaginal has been banished from consciousness it dwells in the unconscious, disguised as childhood memories.

> Archetypal fantasy is the most threatening activity of the human soul as we now conceive it, for our Western rational tradition has placed this activity in the ontologically inferior, the primitive amoral realm of actual childhood.
>
> (Hillman 1975: 23)

Both the 'the child' and 'the archaic' are ways of disguising the realm of the imaginal from ego-consciousness that seeks to defend against the quest into the 'ontologically inferior'. Hillman offered a significant insight that can further illuminate the above elaborations: 'Could childhood be called by its true name – the realm of archetypal reminiscence – then we would not have to become unconscious to find the mythical' (ibid.: 22). The mythical is not necessarily contained in the unconscious. If we consciously accept the inferior child as well as the primitive/archaic as vital elements of our existence – an existence that extends beyond the threats posed by Western thinking – then the mythical could reclaim its rightful place. In light of this assertion the proposed interpretation of Plessner's mythical consciousness becomes an appropriate concept.

We return full-circle to Jung's words:

> Yet every descent is followed by an ascent; the vanishing shapes are shaped anew, and a truth is valid in the end only if it suffers change and bears new witness in new images, in new tongues, like a new wine that is put into new bottles.
>
> (Jung 1911–12/1952: par. 553)

The descent to the archaic parts of our psyche is the way to reconnect with our original nature of mythical consciousness, which is free of eschatological connotations. This seemingly backwards movement will carry us forwards, or even upwards – the direction implied by the ascent – towards the future, where the new wine awaits. The way to steer away from the seeming inevitability of psychic death is to revisit the archaic nature and reconnect with it.

Bishop suggested that analytical psychology can offer the means of attempting such an undertaking. Segal as well spoke of this approach as the antidote 'to the spiritual problems besetting moderns' since it enables a reconnection to the 'state of "primitive" peoples' (Segal 2007: 671). Jung's method carries us further than simply accessing the old part of the psyche, since it also allows for its transcendence. The archaic part of the psyche is brought into the present consciousness and it is carried to the future, where it assimilates with the current (i.e. 'modern') state of the psyche. The 'archaic man' can thus be viewed as a *temporal metaphor* employed by Jung for the advocation of the merits of analytical psychology.

Synchronicity: reconnecting with the archaic

We have examined how synchronicity is founded at the interconnection between the outer (physical/ exterior world) and inner (psyche); it is a manifestation of the psychoid reality. In humanity's primitive thought the inner and outer world are united. Via synchronicity we observe the disappearance of the artificial distinction between human and nature. The principle of synchronicity can thus be seen as the manifestation of the transcendence of the archaic and the modern.

Bishop indicated that 'at significant moments in *Archaic Man*, Jung gestures towards the concept of synchronicity, particularly in his discussion of the difference between the "archaic" and the "modern" worldviews in respect of their attitudes to *chance*' (Bishop 2008: 5130). For the archaic man, meaningful coincidences are part of his everyday experience in the world. This is certainly far from modern everyday attribution of meaning. Justifiably, Plessner wondered:

> Can [man] cease to ask after meaning, aim, or purpose, without confronting death as nothingness? Are we reduced to the alternative of a renewal of Christian transcendence – that is, the eschatological picture of time – or of the flight from time into a mythical view of eternal recurrence?
>
> (Plessner 1957: 245)

If we rely solely on linear temporality then the quest for meaning remains unfulfilled; death will always be casting its shadow over our 'time'. In response to Plessner's pondering, indeed under such temporality meaning would remain entangled with confrontations of 'the end'. The cause of my present state is my past and the cause of my death is my life. Such a stand, dominated by causality, is bound to restrict our experience to eschatological consciousness. Life is an obligatory forward march, with the past as a starting point and the future as the finishing line. Causality is the *de facto* mechanism for such approach.

If, however, we embrace the existence of mythical consciousness, teleology's function is manifested. For Jung, teleology's aim 'is to elicit from the unconscious product a meaning that relates to the subject's future attitude' (Jung 1921: par. 702). Examined from a temporally-linear perspective, teleology could be interpreted as future-meaning-making, and can thus be easily mistaken as an attitude that allows for the prediction of the future – hence Freud's abhorrence of 'future-telling' dreams. If, however, we allow for mythical time to enter consciousness – the time of 'once upon a time' – meaning can claim the central role it was destined for in Jung's theory. Meaning is not attached to the present moment, or to a particular future outcome. It encompasses *simultaneously all temporal demarcations*: the 'now', the 'before', the 'then', the 'modern', the 'archaic', the 'child'. In instances of meaningful coincidences such temporal coexistence becomes evident to the observer.

Jung suggested that 'dreams radiate from a centre, and are only later subjected to the influence of our time. In the final analysis, they are arranged around a *centre of meaning*' (Jung 2008: 10). I would extend this thought so to better understand meaning-making once mythical consciousness is endorsed: our nature is arranged around a centre of meaning but in order for us to conceptualize our lives we subject it to the influence of linear time. For instance, we need to employ linear time in order to narrate a dream, which is in essence timeless and with no particular order. Similarly, we need to delineate our overall experience into past-present-future in order for it to make 'sense'. We could remain attached to a consciousness ruled by linear temporality where explanations are derived mostly by causality and temporal sequencing.

If, however, we were to allow for the heroic descent into the depths where the primordial shadows of our modern psyche rule then a timeless archaic reality would be unveiled. Synchronicity would become once again part of our heritage, and it would invest our existence with meaning. Ascending to the archaic parts of our psyche would also imply acceptance of the interconnection that exists in nature at large, the reality of *unus mundus*. And that would not surprise us, for nature, according to the archaic mentality, extends within as much as outside. We would be enclosed by meaning, which would be effortlessly observed and experienced. Perhaps, this is the glorious ascent that Jung symbolically described.

Notes

1 Borges 1962: 100.
2 Even though Jung's conceptualizations on death are not as widely investigated as Freud's, it was Jung who in *The Psychology of the Unconscious* (Jung 1917/1926/1943) first introduced the notion that the fear of death can lead to neurosis (see Williams 1958).
3 'The cross … is [the hero's] wholeness … His wholeness implies a tremendous tension of opposites paradoxically at one with themselves, as in the cross, their most perfect symbol' (Jung 1911–12/1952: par. 303).

Chapter 6

Time and spatial metaphors

Time is the longest distance between two places.

Tennessee Williams, *The Glass Menagerie*[1]

The endeavour to investigate time regularly and perhaps inevitably involves the spatial dimension. The paradox, however, is that the two concepts refer to different aspects of our experience. As it has been humorously put:

> We can move in space exactly as we please – or stay seated if we prefer. Time, by contrast, appears to carry us along with it. The next morning dawns whether we like it or not, and once it is there, there is no going back. We don't get any younger.
>
> (Klein 2006: 231)

There are essential epistemological differences relating to each concept, the most striking being that contrary to the concept of space 'meditation on the nature of time began with radical doubt about its very existence … Time is far more elusive than space and far more bound up with introspective experience' (Capek 1976: xxvi). And yet, there appears to be no way of talking about time without mentioning space – a case in point is Capek's 'introspective experience', a term that involves the spatial dimension. Most likely this is due to the difficulty we experience in 'getting a hold of' time – literally as well as metaphorically – a difficulty that does not exist with the concept of space. Talking about space without mentioning time is feasible. Whereas the term 'spatialization of time' is consistently found in studies of temporality, the reversed, i.e. 'temporalization of space', is rarely mentioned.

We frequently come across the observed reliance of time upon spatial metaphors in psychoanalytic writings. Some authors consider the interconnection of space and time problematic while others seem to unequivocally rely on it. The former group of writers consider the spatial dimension to hinder explorations of time in the psyche; the fusion of the two concepts, they argue, is deceptive.

A well developed argument was presented by Stephen Mitchell, who identified the use of two different models of writing about the self. The first derives from a spatial metaphor: it views the self as layered, singular and continuous. The second mode is founded on a temporal metaphor and considers the self as multiple and discontinuous. Mitchell argued that the temporal metaphor is more appropriate for the description of the self:

> Thinking about self in terms of time rather than space provides a more useful way of approaching the important issue that the search for the core of the self was meant to solve – the need to distinguish among degrees of authenticity in experience … [T]he self moves in time rather than exists in space, it has no fixed core … The pursuit of an invariant core or true self entails a removal of self from time, a misguided effort to make the shifting organizations of the self experience static.
>
> (Mitchell 1993: 130)

'Authenticity in experience' should replace the spatial metaphors in psychoanalysis, since a multilayered approach allows for temporal unfolding. By adopting a temporality-based language we no longer search for specific elements that are true versus those which are false, but rather each element can be examined as it occurs in time. What is currently true or authentic can later be false or inauthentic.

Others have suggested that spatial metaphors in psychoanalysis should be replaced because they are overly well established and excessively used.[2] Thoughts and feelings, they argue, are not located somewhere. The mind's boundaries are conceptual, rather than spatial, and should be treated as such. We should resist the use of spatial metaphors, particularly of 'inside' and 'outside' – which are 'powerful' and 'familiar' but at the same time 'unproductive, and dispensable notions of location' (Schafer 1972: 422) – and use instead non-spatial language which can be more descriptive, and thus more accurate. Beliefs develop over time; they are dynamic and can simultaneously encompass material from different stages of one's life. Spatial metaphors have also been accused of being inadequate for the description of the therapeutic encounter. Siegelman suggested instead the analogy of 'the arts that move in time, and that move us as they move: to music and dance' (Siegelman 1990: 189–90) for the analytic process, which could be encompassed under the umbrella of temporal metaphors that are dynamic and constantly shifting.

Nonetheless, while proposing the replacement of spatial metaphors Siegelman continued to rely on spatial language – she spoke of 'levels of consciousness' and described the therapist's move in the therapeutic space as 'upward' and 'downward' (ibid.: 189–90). Similarly, Ellis, who pleaded for the abandonment of 'spatial separations', did not manage to steer away from a spatial language. By directing our attention to the 'individual's internal world' (Ellis 2008: 153) she positioned temporality upon the spatial stage. Ironically, theorists who have suggested the replacement of spatial metaphors inevitably relied on them for the description of psyche's temporality; indeed they seem to be 'the very stuff of thought' (Schafer 1972: 420).

Freud relied extensively on spatial metaphors for the exploration of the psyche. From the beginning of his career, in *The Neuropsychoses of Defences* (Freud 1894), he employed the metaphor of Plato's cave to demonstrate the unconscious' obscurity.[3] In *The Aetiology of Hysteria* he compared the analyst to an explorer who, while dealing with the unconscious, comes upon an area of ruins in 'a little-known region' (Freud 1896: 192). Freud's *topographical model* can aptly demonstrate his reliance on spatial metaphors for the exploration of the psyche; the description of the model itself is proof. In *New Introductory Lectures on Psychoanalysis* (Freud 1933) he suggested that for the psyche space and time are interchangeable; the psyche at an unconscious/dream level uses spatial metaphors to express temporal concepts.

For Jung, the distinction between inner and outer is characteristic of the 'modern man', a division that does not exist in the 'primitive' mind. The holistic approach to nature, which Jung searched for, is akin to the primitive state of the psyche, where there is no separation between 'inner' and 'outer'. Our (inner) psyche is one and the same with the (outer) world. This is the return to nature proposed by Jung, the return to the archaic state. However, even though Jung assumed a unified reality, he relied on spatial metaphors for its description.

Beyond the commonplace use of spatial language (i.e. in his referrals to our 'inner' versus 'outer' reality/world) Jung used spatial metaphors in his typology where he distinguished between 'introversion' and 'extraversion'. The metaphor of 'inner space' is also ingrained in the concept of *temenos* (*τέμενος*). Temenos was a sanctified protected space whose 'sacredness was assured by having priests designate a boundary around it – a charmed or protective circle' (Siegelman 1990: 176). For Jung, '[Temenos] is a symbol which is one of the most important motifs in the objectivation of unconscious images. It is a means of protecting the centre of the personality from being drawn out and from being influenced from outside' (Jung 1935: par. 410). By using the image of an isolated space he separated the innermost personality from the outside world.

Searching for the application of spatial language in analytical psychology, 'depth psychology' seems to be the spatial metaphor *par excellence*. This finds resonance in Jung's writings:

> The symbols of the self arise in the *depths of the body* [italics added] and they express its materiality every bit as much as the structure of the perceiving consciousness ... The deeper 'layers' of the psyche lose their individual uniqueness as they retreat farther and farther into darkness. 'Lower down,' that is to say as they approach the autonomous functional systems, they become increasingly collective until they are universalized as extinguished in the body's materiality i.e., in chemical substances ... Hence 'at bottom' the psyche is simply 'world' ... The more archaic and 'deeper,' that is the more *physiological*, [original italics] the symbol is, the more collective and universal, the more 'material' it is.
>
> (Jung 1940: par. 291)

Even though he widely employed spatial language Jung was aware of his use of metaphors for the description of something elusive, which 'can only be realized approximately' (ibid.). Hence the quotation marks accompanying the use of spatial words and phrases through which he emphasized, as well as contained (metaphorically and literally), the spatial metaphors.

Mitchell believed that theories that rely on spatial metaphors view the self as layered and thus supported the abandonment of such an approach: 'It is a mistake to think of one form of experience as more basic or deeper, because experiences are not layered in space; rather they shift back and forth as forms of self-organization over time' (Mitchell 1993: 140). Jung's understanding as illustrated in the above excerpt seems to be in accordance with Mitchell's approach. In Jung's theory – even though, or rather, at the same time that there is mindful use of spatial metaphors, as demonstrated by the use of quotation marks – there is a fundamental shift that is assumed to occur *in time*. Collective symbols that are 'archaic' and 'child-like,' and therefore are believed to have existed at a chronologically earlier stage, become conscious and individual with development (i.e. with the passage of time). The use of spatial metaphors facilitates the portrayal of the process whereupon the symbols 'move' into consciousness.

Jung's search for meaning: beyond space-time duality

At the same time, while using powerful spatial metaphors, Jung employed important temporal metaphors, such as the 'archaic' and 'the child'. The archaic refers to a type of thinking rather than to a chronological description. Likewise, the child does not refer to a chronologically earlier epoch in one's life.

> The empirical idea 'child' is only the means (and not the only one) by which to express a psychic fact that cannot be formulated more exactly. Hence by the same token the mythological idea of the child is emphatically not a copy of the empirical child but a *symbol* clearly recognizable as such … and not – this is the point – a human child.
>
> (Jung 1934/1954: par. 273*n*)

Furthermore, the child is not 'a picture of certain forgotten things in our childhood' but rather it 'represents the preconscious, childhood aspect of the collective psyche' (Jung 1940: par. 273). It is a temporal metaphor that refers to earlier stages of humanity which 'comes into conflict with its childhood conditions, that is, with its original, unconscious, and instinctive state' (ibid.: par. 275). 'The child' and 'the archaic' have temporality ingrained in them. Both motifs suggest an earlier stage of development, i.e. an earlier point in time: in the case of the child, earlier than the grown-up and mature personality; in the case of the archaic, earlier than the modern and 'civilized'.

However, even though both metaphors refer to an earlier state at the same time they *transcend* temporality.

> The child motif represents not only something that existed in the distant past but also something that exists *now*; that is to say, it is not just a vestige but a system functioning in the present whose purpose is to compensate or correct, in a meaningful manner, the inevitable one-sidedness and extravagances of the conscious mind. Our differentiated consciousness [as opposed to the archaic undifferentiated state] is in continual danger of being uprooted; hence it needs compensation through the still existing state of childhood.
>
> (Jung 1950: par. 276)

Like the archaic, the child motif compensates our tendency to move away from our nature, from the original state of our psyche. Thus, through these temporal metaphors we remain connected to the relative temporality of the psyche. They manifest its transcendental quality. There is multiplicity and opposition within the psyche that unfolds throughout our life.

It is therefore inaccurate on Mitchell's behalf to suggest that Jung viewed the self in 'a singular, integrated sense' (Mitchell 1993: 108[4]). Mitchell focused exclusively on the aspect of the self that contains inborn archetypal potentials and overlooked a vital part of Jung's theory, namely the *inherent duality* in the psyche that results in constant opposition. Jung's psychology regards the self as non-static, and as operating in time. This is in accordance with Mitchell's advocation for the use of temporal metaphors: 'The pursuit of an invariant core or true self entails a removal of self from time, a misguided effort to make the shifting organizations of the self experience static' (ibid.: 130).

There is a fundamental sense of development in Jung's conceptualizations.

> Some [analytical psychologists] tend to define the self as the original state of organismic integration. Others see it as an image of a supraordinate unifying principle. Both groups make use of Jung's frequent references to the individual personality as 'emerging from' the archetypal potentials contained within the self.
>
> (Samuels *et al.* 2003: 137)

In the case of the former group of theorists, the self continuously develops from deintegration to reintegration. In the case of the latter group, the self encompasses and unites under its auspices all the opposing and conflicting elements of the psyche. In both cases there is continual movement, embedded in the notion of emergence: 'The picture [of the conception of the self] is not one of static "order" but rather a dynamic integration' (Samuels 1985: 90).

A dichotomous understanding of spatial metaphors as implying a static and singular state versus temporal ones as representing a multiple and discontinuous perspective seems rigid and imprecise. Also, a spatial metaphor does not necessarily entail stillness; spatiality does not exclude movement.[5] Even though many theorists disagree with the extended use of spatial metaphors in psychoanalysis, others take for granted the unification of space and time and use a language that incorporates both. Space and time are synchronized, as it were, to create a joint, dynamic field upon which the analytic situation unfolds.[6] It has been indicated that even the use of blatant spatial metaphors – such as the distinction between internal and external worlds – implies temporality:

> [W]henever there is spatial movement, there must also be temporal movement. In nature the passage from one space to another involves time. The two spaces will always exist; only time can bring them into contact. If we formulate the problem of the internal world and external reality in terms of a spatiotemporal metaphor, we see how these two perspectives in the psychoanalytic world can be united.
>
> (Brizzi 1998: 66–7)

Brizzi's observation regarding the *simultaneous* existence of a spatiotemporal metaphor takes Mitchell's call for the unification of the two distinct metaphors a step further.

In light of the above, the request for the replacement of the spatial metaphor seems, at best, far-fetched. It even appears to be an indefensible quest. Why should we strive to replace or overshadow the use of spatial metaphors (with temporal or with any other sort of metaphors or concepts) when they have diachronically proven to be a very powerful tool of human thought? Perhaps it is not the use of the temporal metaphors that is misleading but rather the search for their substitution and the unfounded need for dichotomy. A more balanced approach is in order where temporal and spatial metaphors have equal space (pun intended) and gravity. Conceivably, even, a third element should be allowed to hold the joining of the two, much in the same way as Jung conceptualized the third to be larger than the elements that created it.

In the Pleroma time and space, even though they exist as a pair of opposites, cancel each other out. Humans, on the other hand, unaware of the actual negation of opposites and of the perfect balance that exists in the Pleromatic stratum, fall victim to the seeming opposition. In order to be released from these conflicting energies we must learn to distinguish ourselves from them: 'If we know how to know ourselves as being apart from the pairs of opposites, then we have attained salvation' (Jung, as cited in Hoeller 1982: 47–8). Having in mind the Gnostic cosmology as narrated by Basilides we could conceptualize yet another application of the unification of space and time. A possible way of resolving the tension between these forces is to allow for the opposition to exist and embrace the duality.

Ideally, temporal and spatial metaphors should coexist. In Jung's opus there is constructive dialectic between the temporal (constant movement in time, existence of conflictual/opposing elements within the self) and the spatial (a singular and continuous self that lies at the core/depth of the psyche, the archetypal original self). These two elements appear to be united by the search for meaning. Meaning surpasses both the temporal and the spatial categories of the psyche:

> [T]he 'archaic' and the 'primordial' are not necessarily chronological categories, but can also signify an intensity or authenticity of experience. This synchronic, as well as diachronic (or chronological), dimension to Jung's understanding of the 'archaic' points to the link between the 'archaic' and the concept of synchronicity, the latter being Jung's way of developing a discourse centred on the category of meaning.
>
> (Bishop 2008: 515)

Even the archaic and the primordial – which are justifiably regarded as temporal metaphors due to their etymology[7] – are beyond time. Rather, they are not only about a temporal quality but *also* about a quality that 'measures' their meaning. *Meaning is their juncture.* This could also apply to the principle of synchronicity. Even though the term etymologically implies a temporal basis for the theory, it is not about time, but rather about its transcendence. Synchronicity is about *time's demolition on the altar of meaning*.

Hillman described the world of the child as '*systematically* haphazard, afloat; all things held together by spontaneous amazing synchronicities' (Hillman 1975: 23). The child exists in a world where duality and opposition dissolve into synchronistic occurrences. The oxymoron of 'systematically haphazard' experienced through the standpoint of the child, results in a natural state of being where 'all things intending the child, pointing to him, keeping him alive in the centre of importance, even while he does nothing, wills nothing, knows nothing' (ibid.). The worldview of primitive people, as described by several anthropologists as well as Jung, is characterized by such an understanding. The world is entrenched by meaning that defies causality. The archaic/primitive, like the child, effortlessly exists in a world deep-rooted in meaning.

Mitchell suggested that, 'The apparent contradiction [between the spatial and the temporal approaches to self] provides a creative tension that makes possible a more complex and useful view of the place of self within experience and the transformations of self in psychoanalysis' (Mitchell 1993: 102). Jung's search for meaning, which lies beyond space and time, can be viewed as the creative outcome of such tension. Time and space as opposing elements are the vital catalysts for the manifestation of such outcome. The principle of synchronicity and its inherent connection to meaning is a demonstration of such creative effect. *The psyche* – represented by the spatial metaphor of interiority – and *the world* – found in the opposite spatial realm, that of the exterior – convene to create a meaningful connection which is not about the chronological occurrence of the events; it is, nonetheless, dependent on, and created out of its *occurrence in time*.

Notes

1 *The Glass Menagerie*, Part 4, Scene 7.
2 See for example Loewald 1962; Schafer 1972; Frank 1996; Meissner 2007.
3 Noel-Smith 2002.
4 'This experience of self as integral and continuous also has been very much illuminated within contemporary psychoanalytic thought … Among the antecedents of this approach is … Carl Jung's notion of an archetypal predisposition for people to form a singular, integrated sense of self' (Mitchell 1993: 107–8).
5 See Slochower 1999; Crastnopol 2002.
6 See Brizzi 1998; Sabbadini 1989.
7 *Archaic*, from the Greek *αρχή* (arche) for 'beginning'; *primordial*, from the Latin *primordium* for 'origin'.

Concluding remarks

> [T]he Hatter … [tossed] his head contemptuously. 'I dare say you never even spoke to Time!'
>
> 'Perhaps not,' Alice cautiously replied; 'but I know I have to beat time when I learn music.'
>
> 'Ah! That accounts for it,' said the Hatter. 'He won't stand beating. Now, if you only kept on good terms with him, he'd do almost anything you liked with the clock.'
>
> Lewis Carroll, *Alice's Adventures in Wonderland*[1]

Jung's theory attempts to grasp the nature of the individual psychological world in the context of the wider universe. Jung conceptualized reality as encompassing the psychic as well as the physical dimension: the psyche exists above and beyond the physical reality; it is a phenomenon in its own right which transcends the material world.

The timeless unconscious was Freud's original contribution to the understanding of time in the psyche; it was an unprecedented discovery (Green 2002: 162). Reyner and Tuckett (commenting on Matte-Blanco's writings) observed that the field of psychoanalysis has not 'taken seriously enough the idea that the mind works within a framework of timelessness and spacelessness … Analysts have tended not to consider the matter or have moved their discipline towards conventional psychology and conventional logic' (Reyner and Tuckett 2005: 5). Via Jung's 'time theory' we can detect a turn towards the original essence of psychoanalysis. Through his understanding of time Jung has moved away from 'conventional logic' that relies heavily on cause and effect – an essential feature of his contemporary mainstream scientific approach – and into the realm of the acausal.

In distinguishing the vital differences between Freud's and Jung's understanding of time, we realize that Jung's adherence to relative time is in accord with one of Freud's 'most profound discoveries' (Freud 1933: 74) namely timelessness; Freud's 'unprecedented discovery' of the timelessness unconscious was in fact developed by Jung's 'time theory'. By relying on relative temporality, Jung moved beyond Freud's dichotomous hypothesis and widened the temporal perspective. The emergence of consciousness did not create an opposing 'kind' of time. Jung incorporated Freud's observation and took timelessness a step further; time's relativity was the next step to timelessness. *In the psyche everything is simultaneous and contemporaneous.*

For Jung, the past and the future are of equal importance. The past anchors our experiences; the direction of the psyche points forwards, into the future. At the same time, the temporal relativity that lies at the foundation of his 'time theory' assumes multidimensionality. The future does not refer to a particular fixed point in time as expressed by linear demarcations; there is a *future dimension* towards which the psyche gravitates. Understanding teleology in this light is in accordance with Jung's overall 'time theory' which is characterized by overarching relativity. There is no pre-fixed *telos* that could be foretold; the past does not cause the future. In relative time, temporal demarcations are equated; time's dimensions co-exist. The future holds as much power as the past, as well as the present moment. And at the same time they are annihilated.

In consequence, referring to teleology, as well as foreknowledge, is inaccurate, unless we supplement the terms with an *underlying relative temporality*. The psyche is characterized by teleological directionality instead of a future 'target point'. It has a vision of its possible wholeness. Wholeness can be accessed at any given moment of our life – as demonstrated by children's dreams – because it is ever-present. It is relative and thus it transcends linear time and causality. According to the psyche's relative time, the future, like the past, is within our reach not by accepting para-scientific explanations, but rather, by widening our scope to include an unorthodox temporal reality.

It seems that the way we conceive time affects our experience of it. Cases in point is Freud's adherence to the containment of timelessness in the unconscious, which led him to dismiss the possibility of prophetic dreams; he also admitted being unable to experience the oceanic feeling. On the other hand Jung's acceptance of time's relativity allowed for his exploration of 'prophetic' dreams and synchronistic experiences. Through his hypotheses, Jung encouraged his reader to develop an approach to time that would invite the investigation of non-linear, multidimensional temporal experiences.

Noel-Smith, in an attempt to negotiate the psyche's paradoxical temporal expressions, suggested the creation of 'temporal thought':

> If we are lucky enough to internalize securely and integrate the functions of containment and insight, of inner space and inner time, we can contemplate the ideas of eternity and infinity without fear. In this integration, and the acknowledgement of the loss of the idea of omniscience, we can create our own temporal and spatial boundaries which take account of our time and our space in the real world. I suggest that it is temporal thought which creates the possibility of drawing those boundaries and it is with time that we circumscribe our world.
>
> (Noel-Smith 2002: 435)

We live our lives in a clock-bound world dominated by its fixed keeping of time. Nonetheless, assuming we are 'lucky enough' to have the 'luxury of time' so as to *pay attention*, we might realize that we are also inundated by paradoxical temporal occurrences in dreams, in the analytic situation, or even in inexplicable – by causality – coincidences that defy the clock. As Marcel Proust, who had exhaustedly searched for *Lost Time*, affirmed: 'The time which we have at our disposal is elastic; the passions that we feel expand it.'[2]

Relying on luck is not an adequate approach. However impossible it is for consciousness to grasp concepts that lie beyond conventional time, a sufficient way of negotiating our temporal experience is imperative. We should aim at cultivating 'temporal thought'. As Noel-Smith accurately observed, 'time and space are things that we do, and have to do, to make sense of a world which really makes no sense at all' (Noel-Smith 2002: 435). I would add that sense-making could be facilitated if we, following Jung's lead, search outside the realm of the ordinary and temporally-linear.

Time and synchronicity: manifesting meaning

The exploration of the physical spectrum via physics was a necessity for Jung's theory since psychological knowledge alone did not adequately explain his observations. 'The psychoid nature of the archetype contains very much more than can be included in a psychological explanation' (Jung 1958: par. 852). Jung moved away from causality in an attempt to extract meaning: the acausal connecting principle of synchronicity could thus be regarded as a manifestation of his meaning-making quest. During synchronistic occurrences the psychoid's inherent yet transcendent meaning is observed. Time is the crucial catalyst for these occurrences and can provide the missing link between these two fundamental concepts: synchronicity is *a temporally-bound expression* of the psychoid's transcendent nature.

This manifests when we appreciate that Jung's adherence to relative temporality does not assume a time-less vacuum. A time-bound dimension is necessary for the observation of temporally relative reality. Consciousness, being enthralled by fixed time, creates differentiation and thus temporal demarcations. Paradoxically, there appears to be an imperative function for such ostensible limitation. It is only in consciousness that the otherwise latent Meaning can be 'captured'; the psychoid's transcendent all-encompassing reality can be experienced in consciousness-bound time.

Synchronicity is the observation of meaningful connections existing regardless of causality and temporal linearity. Such connections are observed *in time*, i.e. in the temporal parameters within which an observer consciously operates. A dream which had 'predicted' a month ago today's events, even though it occurred while the ego was dominated by timelessness, was nonetheless brought to awareness upon consciously recalling it. The *attention bestowed* upon the dream in conscious time is the reason (i.e. the underlying cause) for the occurrence of synchronicity.

A distinction should be made between different 'levels', or perhaps dimensions, of meaning. During synchronistic occurrences the conscious observer gets a glimpse of the all-encompassing-yet-concealed Meaning that exists in nature at large, whereupon he attributes 'meaning'. Time's emphasis in synchronicity (syn + chronos) should not be considered a limitation, but rather a deliberately selected yet underdeveloped stipulation of the principle.

The ripe time for knowledge

Centuries ago Saint Augustine, in God-fearing spirit – for such assertions could be conceived as blasphemous – posed a hypothesis that is strikingly similar to Jung's:

> I am investigating, Father, not making assertions. By God, protect me and rule me ... [Does past, present and future] exist in the sense that, when the present emerges from the future, time comes out of some secret store, and then recedes into some secret place when the past comes out of the present? Where did those who sang prophecies see these events if they do not yet exist? To see what has no existence is impossible. And those who narrate past history would surely not be telling a true story if they did not discern events by their soul's insight. If the past were non-existent, it could not be discerned at all. Therefore both future and past events exist.
>
> (Saint Augustine 1998: 233)

Saint Augustine's *Confessions* were completed in AD 400. Fifteen centuries later Jung would seek the support of modern science, in order not to be misunderstood – not by God, but by men of his time – for suggesting that the future already exists. As Jung wrote in 1945, the work of physicists '[has] given me courage to venture to establish a connection which came to me quite unexpectedly'[3] (Jung 1973: 363). Certainly, courage was in order, for Jung ventured into uncharted territories in order to elaborate his observations. He was to first express them under a pseudonym in a Gnostic poem. We are time-and-space-bound, but live in timelessness and spacelessness, in a world of eternity and infinity. It is as if through *Septem Sermones Ad Mortuos* Jung came into contact with a formula on how the universal and the personal realms connect and interplay. These verses served as a doorway to another dimension of Jung's psyche, where the mysteries of time and eternity dwelled. With Gnosticism as his vehicle and Basilides as his voice Jung was able to infiltrate time's complexities.

In *Memories, Dreams, Reflections* Jung addressed temporality's peculiar nature as he came to experience and understand it throughout his life. Future, present and past are mere demarcations of a temporally unified reality; time and space are relative. However, 'unlimited knowledge [which is] present in nature ... can be comprehended by consciousness only when the time is ripe for it' (Jung 1961: 307). In consciousness, which arises at a particular point in time, absolute knowledge manifests. Even though absolute knowledge is accessible, only during a particular point in time 'when the time is ripe for it' is it ready to be voiced and understood. At a particular time-period one reaches beyond fear of God and into his own soul where his truth rests and needs to be expressed. This specific point in time could be conceived of as 'ripe' in respect to one's personal life, as well as to history at large. Saint Augustine's knowledge could only have been voiced in his confessions to God; Jung's observations had to be initially masked by Basilides' voice and later supported by Pauli's scientific knowledge.

Death and time: embodied and disembodied time

Death: the end of time. An attitude pertinently described by Bonaparte: 'For in living our time we die of it' (Bonaparte 1940: 429). Such approach presupposes time to be the destroyer of life. Jorge Luis Borges beautifully described the anxiety associated with such outlook:

> Time is the substance I am made of. Time is a river that grabs me, but I am the river. It is a tiger that rends me, but I am the tiger. It is a fire that

> consumes me, but I am the fire. The world is, unfortunately, real. I am unfortunately, Borges.
>
> (Borges 1974: 771)

Jung, moving away from 'the unfortunately real world' and, operating from a symbolic point of view, assumed that death holds a vital role in the psyche's revivification – 'vital' not only as 'crucial', but also as 'imbued with life'. He attested: 'A man should be able to say he has done his best to form a conception of life after death, or to create some image of it – even if he must confess his failure' (Jung 1961: 302). Having the Jungian paradigm in mind, I would suggest that creating an image of death is in fact the quintessential treatment of death.

Physical demise is indeed a certainty; however, so is birth. Birth is as perplexing and important as death. It is death's prerequisite (we must be born in order to die). Also, as many religions/systems express, birth is death's outcome; upon death there is re-birth. Birth and death presuppose one another. They are in fact the epitome of the necessary pairs of opposites. The ability to comprehend this paradoxical reality is what grants us our 'deep and significant identity' (Spengler 1991: 89).

Assuming that death casts an overbearing shadow over the psyche – a shadow that prevents us from gazing forward and experiencing the potentially regenerating aspect of death – seems to be an incomplete take on reality. This assumption is based on an embodied sense of time: time as it connects to our biology, to our physical life and demise. Jung's approach, on the other hand, embraces *disembodied time.* It seeks for explanations beyond causality, and commonsensical time. His adoption of a symbolic view allowed for his theory's expansion beyond anxiety of physical death. On the symbolic plane, death is a psychic experience pregnant with hope and vitality. By embracing the non-consequential, Jung made room for the paradoxes of the psyche to unfold.

Perhaps because our temporal experience is inseparable from our experience as living beings it is exceedingly difficult to perceive time as extending beyond bodily death. How can we, while still living, comprehend or even fathom a life-less state? A rough approximation is the best-case scenario. This is why the employment of the symbolic is of vital importance. To extend this conceptualization it would be useful to further explore the metaphor of 'the living' and 'the dead'.

Robert Pogue Harrison, using a fitting example from Homer's *Odyssey*, offered this: '[Odysseus'] history cannot go forward in Kalypso's earthly paradise, for the latter is outside both world and time' (Harrison 2008: 7). Life unfolds in time; and, consequently, it does not and cannot take place outside it. In paradise, where there is no death or birth, where nothing perishes and all things flourish *ad infinitum* time is nonexistent.

Death does not necessarily imply the end of time; it merely entails the end of time as perceived by the living. Harrison evoked the story of Odysseus' descent to Hades, where his dead companions enquired about the life and the living they left behind: 'No resolution is final, of course, and even death does not put an end to certain cares (as Odysseus learns when he talks to the shades of his dead companions in the land of the dead)' (ibid.). These circumstances bear a resemblance to Basilides' visit by the dead who were in search of what they did not find in Jerusalem.[4] Also, in *Memories, Dreams, Reflections* (in a chapter entitled 'On Life after Death') Jung wrote of the spirits of the dead who would visit the living so as to gain information. Jung realized that:

> Apparently ... the souls of the dead 'know' only what they knew at the moment of death and nothing beyond that. Hence their endeavor to penetrate into life in order to share in the knowledge of men. I frequently have a feeling that they are standing directly behind us, waiting to hear what answer we will give to them, and what answer to destiny. It seems to me as if they were dependent on the living for receiving answers to their questions, that is, on those who have survived them and exist in a world of change.
>
> (Jung 1961: 308)

'The dead' are amongst 'the living'; or, in the case of Odysseus, the living are amongst the dead. What separates the living from the dead is that the living dwell 'in a world of change'. Change defines our time-bound existence, ruled by linear temporality and marked by chronological, ever-changing circumstances. The 'time of the living' in Jung's opus signifies the time of consciousness. Change can only occur in conscious time, when it can be observed. This is the time within which we become witnesses to synchronistic occurrences. The spirits of the dead do not posses knowledge beyond what they acquired while still alive, that is, incarnated in a time-and-space-bound body. 'The dead' are in a diffused state: body-less, age-less. Perhaps, then, the 'time of the dead' is equivalent to the change-less, relative temporality that dominates nature at large; it is the Pleromatic eternity. In consciousness (or, in the 'time of the living') there is no permanence. This is temporality as expressed in Creatura: things continue to evolve, evolution is perceived, questions can still be answered, and knowledge is to be acquired.

Jung concluded:

> The mind of the living appears, therefore, to hold an advantage over that of the dead in at least one point: in the capacity for attaining clear and decisive cognitions. As I see it, the three dimensional world in time and space is like a system of co-ordinates . . . necessary if any distinction or discrete contents is to be possible. Any such operation seems to us unthinkable in a state of diffused omniscience, or, as the case may be, of subjectless consciousness, with no spatio-temporal demarcations. Cognition, like generation, presupposes an opposition, a here and there, an above and below, a before and after.
>
> (Jung 1961: 308)

Our space-and-time bound existence that is the cause of our most profound anxiety is, at the same time, the reason we posses cognition of any kind (be it reasoning, awareness, perception or judgement). Similarly, Basilides spoke of differentiation as a precondition of human existence. We cannot conceive meaning unless it is embedded in opposition created by temporal demarcations, in 'befores' and 'afters,' in 'the living' and 'the dead'.

However, to equate consciousness with temporal demarcations and parameters which in essence *require* linearity (a 'before' will cause an 'after', 'life' will cause 'death') seems to be far from the whole story. Linear thinking is compulsory; at the same time it is an inadequate approach. An apt example is this book: linearity is responsible for the frustration experienced while attempting to explore temporality via a piece of writing that is supposed to be characterized by logical connections. In attempting to write about time I am confined within the lettering; or, more accurately, the meaning is bound to be confined to the lines, for writing is obliged to obey linearity. Presumably, every effort to explore the matter gets lost in translation, in the cognitive boundaries imposed by my fixed temporal and spatial co-ordinates.

In 'Burnt Norton' (Four Quartets), while skilfully conveying our entanglement in conscious time, T. S. Eliot offers an expanded view of consciousness.

> Time past and time future
> Allow but a little consciousness.
> To be conscious is not to be in time
> But only in time can the moment in the rose-garden,
> The moment in the arbour where the rain beat,
> The moment in the draughty church at smokefall
> Be remembered; involved with past and future.
> Only through time time is conquered.
>
> (T. S. Eliot 1944: 5–6)

Only *in time*, the poet observed, can the glory of the moment unfold. We cannot be conscious unless we are in the present moment. Time, however, is not equivalent to consciousness; 'to be conscious is not to be in time'. T. S. Eliot's words echo Jung's understanding: time is not contained in linearity; it is relative.

Jung's theory could take this poetic evaluation a step further. Even though we are bound by time, time is not restricted to our sense-bound observations, 'to the moment in the arbour where the rain beat'. As Harrison noted, certain cares are not resolved by death and, as a consequence, they are not resolved by our physical, living experience of time. There is a sense of time beyond eschatological temporality, of which physical death is considered its culmination. 'The dead' come to 'the living' for they lack consciousness and, therefore, cognition. Diffused in space and time they have no access to earthly goods.

There is a *multidimensional reality* that engrosses all that is and has been, as well as what shall be, and concurrently reaches beyond life and death into eternity and infinity. As Jung's theory suggests, the psyche's timelessness is not contained solely by the unconscious. The relative dimension of time can be experienced and observed consciously, regardless of its mysterious nature. This can be achieved by descending into mythical consciousness.

Symbols, spatial metaphors and the poetry of time

Symbolic language, which is regularly employed in myths, appears to be the most suitable medium of expressing an otherwise incomprehensible experience. Perhaps this is the reason why so many poets have dealt with the manner in which time is felt and perceived – and most likely it is why I find myself frequently quoting such poems, as they can be the most suitable medium through which these experiences can be voiced. Employing symbols, that is, attributing an image to an elusive concept, creates a concrete dimension. In this sense, *every symbol is a spatial metaphor*. Conceivably, this is why spatial metaphors are imperative for the conceptualization of temporality.

I have proposed that the symbol of the Uroboros can make temporality of the *unus mundus* palpable. The Uroboros contains the opposition of time-boundness and timelessness. The linearity of consciousness (symbolized by the serpent's vertical presentation) is simultaneously presented as an endless circle (the tail-biting serpent). If we are to deconstruct the psychoid in these terms, we could conceive it as the exemplary spatial metaphor employed to describe a temporal reality. In actuality, the 'inner' and 'outer' are not divided in space, but rather they exist simultaneously. We could assume this to be the function of space in relation to the psyche: only through the tangible parameters of space can unattainable temporality be regarded. Through the punctuation inherent in the spatial experience we can appreciate the psyche's paradoxical temporality. This conjecture makes the observed interrelation of psychic space and time (in psychoanalysis as well as in everyday life) comprehensible.

Yet, the above interpretation weakens the role of relativity in the psyche. The spatialization of time concretizes both time and space. For example, the above interpretation of the Uroboros, even though it contains the temporal opposition of timelessness and time-boundness, is restricted by a linear interpretation. As I have argued, even though seemingly non-linear, a circle is still based on a (curved) line; it is unidimensional. If, however, we are to amplify the symbol of the Uroboros to incorporate temporal and spatial relativity we must expand it beyond cyclicality and thus beyond unidimensionality. The Uroboros is not a circle, but rather a three-dimensional image, i.e. a tale-biting snake. It does not simply contain the paradox of the timeless and the time-bound. It contains the *multidimensional paradox* of the simultaneous existence in time *and* space, of all that was, is and shall be; of all that is timeless and transitory, of all that is above and below; of all that is conscious and unconscious.

As exemplified by *Septem Sermones Ad Mortuos*, the poetic and symbolic language is a suitable medium through which Jung's relative temporality in the psyche could be illustrated. Metaphors and symbols are imperative in expressing not only time in the unconscious, but also temporality in mythical consciousness. For this reason I would suggest a different application of the term 'chronopoietic function'. 'Chronopoietic' is a term coined by De Simone (1997) to signify 'time-making': it derives from chronos [χρόνος] and poietic [ποιητικός] (maker, productive, formative). According to the original meaning of the term, the therapeutic encounter has the capacity to create time. The rhythm and duration of the sessions create a sense of cyclical time that is contrasted with the linear time of the analytical process, which has a set beginning and ending. By formulating a new sense of time within psychoanalysis we reinvent our history. What was previously understood in a certain way, once analyzed within the contrasting temporalities of the psychoanalytic setting, has the potential of being experienced anew. Due to this constant rediscovery of the sense of time De Simone proposed that 'the analytic situation has a *chronopoietic* function' (De Simone 1997: 21). Beyond the productive and formative meaning of 'poietic' [ποιητικός] it is also the word for poetic. We could thus widen the characterization to include not just the making of time, but also *the poetry of time*. There is a chronopoietic function in our human experience at large: life invites the creation of poetic imagery through which our temporal experience can be portrayed.

Saint Augustine, in an often-quoted observation, asserted:

> Who can comprehend [time] even in thought so as to articulate the answer in words? Yet what do we speak of, in our familiar everyday conversation, more than of time? We surely know what we mean when we speak of it. We also know what is meant when we hear someone else talk about it. What then is time? Provided that no one asks me, I know. If I want to explain it to an inquirer, I do not know.
>
> (Saint Augustine 1998: 230)

The bewilderment in Saint Augustine's mind arose from the confusion between the 'knowing' presupposed in communicating the nature of time (that is, the cognitive aspect of knowledge) and the 'knowing' that we acquire 'internally' that is not restricted by linearly temporal or spatial coordinates. 'Knowing' in order to communicate an idea relies heavily on conventional logic. Paradoxically, however, the 'familiar everyday' sort of temporal experience is not confined to the realm of unadventurous logic. In order for the whole of the temporal experience to be communicated we must include the sense of timelessness that lies outside the apparently explicable. *Time requires timelessness*: 'there's … a time for growing and a time for dying … but more than all(as all your more than eyes tell me)there is a time for timelessness' (sic) (E.E. Cummings 1997: 98). Only by allowing, experiencing and expressing these polarities in their multidimensionality can the entire temporal spectrum unfold.

Jung spoke of the 'diffuse cloud of cognition' (Jung 1961: 308) that exists outside the three-dimensional world of space and time: 'What is here separated into ordinates and abscissae may appear "there," in space-timelessness,[5] as a *primordial image* with many aspects, perhaps as a diffuse cloud of cognition surrounding an archetype' (ibid.). Unless we are able to communicate the 'other side of time' (notice the use of yet another spatial metaphor) we cannot wholly explain the concept. In the cloud of cognition, 'here' is 'there' and 'everywhere' and 'before' can be 'now' as well as 'after'. Once timelessness with all its irregularities is embraced we could, perhaps, really know Time. Conceivably, even, as the Mad Hatter remarked after asserting to have spoken to Time, 'if you only keep on good terms with [Time], he'd do almost anything you like with the clock' (Carroll 1960: 71).

There is, or rather, there should be, time for timelessness in our lives. Jung slipped the bounds of linearity, he ascended above the cloud of cognition, and traversed the deepest parts of our psyche; he annihilated past and future, even the laws of physics; he dared to transcend death. His opus made this clear: by accepting the acausal we can experience the sense of time that dwells in mythical consciousness. Rather than assuming that time is 'that which denies us the eternality we feel we were meant for but can experience only in unconsciousness' (Pollock 1971: 436) by embracing Jung's attitude we can grasp a different reality. By *consciously experiencing timelessness*, through the symbolic life, we can partake in eternity. In allowing timelessness to come forth we create the space for astonishing experiences and connections; we enrich our lives, inner and outer. Indeed, there must be *time* as well as *space* for timelessness. Through timelessness we create space; in timelessness we find Time.

Notes

1 Carroll 1960: 71.
2 Proust, as cited in Terr 1984: 641.
3 Jung was referring to Pascual Jordan's research on telepathy.
4 The importance that Jung placed on his communication with the dead in *Septem Sermones* was revealed with the publication of *The Red Book*: '[T]he discussions with the dead [in *Septem Sermones ad Mortuos*] formed the prelude to what he would subsequently communicate to the world, and that their content anticipated his later books. "From that time on, the dead have become ever more distinct for me as the voices of the unanswered, unresolved and unredeemed." The questions he was required to answer did not come from the world around him, but from the dead' (Jung 2009: 346 *n*78).
5 Perhaps it would be more accurate if Jung had spoken of 'spacelessness-timelessness' as opposed to 'space-timelessness' (Jung 1961: 308). By keeping the dimension of space fixed partial relativity is implied. However, Jung's words indicate that time *and* space are relative; there is timelessness *as well as* spacelessness. It appears that Jung was using space-time as a unified dimension (in the same way that modern physics refers to space-time), to which he adhered '-lessness' (that is: *space-time*lessness). This interpretation is in accordance with his overall time theory.

Appendix

Freud on memory, the primal scene and the concept of *Nachträglichkeit*

In the earliest phase of his work (in what has been characterized as the affect-trauma frame of reference which lasted up to 1897) Freud believed a single event that had occurred in one's actual life to be the cause of the trauma; trauma was considered to be a 'once-and-for-all affair' (King 2005: 189). The focus of psychotherapy was on remembering the original trauma; by making it available to memory the symptoms would subside. Nonetheless, as Freud had observed, time had a particularly strange behaviour in the psyche. As he observed in 1897 in *Draft M* 'neglect of the characteristics of time are no doubt essential for the distinction between activity in the preconscious and the unconscious' (Freud 1954: 246–8).

During the next phase of his theory Freud suggested that the timeless unconscious contents, which remain active and are unforgettable, seek opportunities to return to consciousness as if no time has elapsed. The aim for psychotherapy is to rid unconscious memories of their unconscious excitation; *forgetting* is the ultimate goal. As he stated in *The Interpretation of Dreams*: '[The task of psychotherapy] is to make it possible for the unconscious processes to be dealt with finally and forgotten' (Freud 1900: 578). Bringing the unconscious – the storehouse of never-dying memories – under the command of consciousness seemed to be the only possible solution to disarm the all-mighty immortal memories of the trauma.

In his later writings Freud elaborated his theory regarding the workings of memory. Memory in the unconscious is based on memory traces that cathect instinctual energy and give rise to the contents of instinctual wishes. In order for an idea to become available (i.e. to enter consciousness) it would have to get attached to a memory residue: '[Residues of memories] were at one time perceptions, and like all mnemic residues they can become conscious again' (Freud 1923: 20). Since none of the contents of the unconscious can expire, nothing is ever forgotten; residues of memory were deemed indestructible. Recalling is achieved as long as one goes back in time: 'Everything is somehow preserved and ... in suitable circumstances (when, for instance, repression goes back far enough) it can once more be brought to light' (Freud 1930: 69). In the following excerpt Freud, who had great admiration for archaeology, used the allegory of Rome's ancient buried ruins to illuminate his position.

> It is hardly necessary to remark that all these remains of ancient Rome are found dovetailed into the jumble of a great metropolis which has grown up in the last few centuries since the Renaissance. There is certainly not a little that is ancient still buried in the soil of the city or beneath its modern buildings. This is the manner in which the past is preserved in historical sites like Rome.
>
> (Freud 1930: 69–70)

The eternally enduring past, even though buried out of sight, is nonetheless preserved intact under the newly erected present. However, in the same passage, he appeared ready to accept the possibility that the contents of mental life can be obliterated and cannot be re-established in consciousness:

> It is always possible that even in the mind some of what is old is effaced or absorbed – whether in the normal course of things or as an exception – to such an extent that it cannot be restored or revivified by any means; or that preservation in general is dependent on certain favourable conditions. It is possible but we know nothing about it. We can only hold fast to the fact that it is rather the rule than the exception for the past to be preserved in mental life.
>
> (Freud 1930: 71)

He spoke of rules regarding the behaviour of past contents. Nonetheless, these rules did not always apply.

Freud made another important observation regarding past contents. New experiences and impressions, as well as memories of old experiences, have the capacity to alter these otherwise durable memories. Every new experience transforms the mnemic content. 'It is generally thought that it is time which makes memory uncertain and indistinct. It is highly probable that there is no question at all of there being any direct function of time in forgetting' (Freud 1901: 274–5). He returned to this in 1907 to add an illuminating thought – which he placed nonetheless in a footnote – dealing explicitly with the idea that repressed materials are unaffected by time.[1]

> In the case of repressed memory-traces it can be demonstrated that they undergo no alteration even in the course of the longest period of time. The unconscious is quite timeless. The most important as well as the strangest characteristic of psychical fixation is that all impressions are preserved, not only in the same form in which they were first received, but also in all the forms which they have adopted in their further developments. This is a state of affairs which cannot be illustrated by comparison with another sphere. Theoretically every earlier state of the mnemic content could thus be restored to memory again, even if its elements have long ago exchanged all their original connections for more recent ones.
>
> (Freud 1901: 274–5*n*)

Repressed memories are not unaffected by the passage of time altogether. This process Freud associated with the concept of *Nachträglichkeit*: once a past experience is recollected in the present, a reorientation in relation to reality takes place.

The most comprehensive and recurring treatment of *Nachträglichkeit* is found in *The Wolf Man Case Study* (Freud 1918) where he linked the concept to the primal scene (i.e. the memory of witnessing an act of parental intercourse). In the original writing of *The Wolf Man*, Freud used *Nachträglichkeit* to suggest that the dream about the wolves represents actual impressions from the primal scene that the patient had at the age of one and a half. According to this first version of the case study, the dream that occurred twenty years later gave rise to the neurosis since the patient was able to grasp retrospectively the meaning of his past experience.

However, (as mentioned in Chapter 2) Jung's theory of regressive fantasies instigated a revision of Freud's understanding of the primal scene and its relation to *Nachträglichkeit*.[2] For Jung, regressive fantasies are the workings of the regressive movement of the libido: 'Only later, when the regressing libido, seizing hold of anything suitable that lay in its path, had turned all this into a fantasy. Then the fantasy, keeping pace with the regressive movement of libido, came back [to the present moment]' (Jung 1913: pars. 394–5). Thus, even though in the original version of *The Wolf Man* case study Freud maintained that the reinterpretation of the past event was responsible for its traumatic meaning, in the revised version he suggested that the traumatic event could have been a *fabrication* of the patient (as opposed to an actual event).

Laplanche and Pontalis in an attempt to explain Freud's theory suggested that for Jung 'reinterpretation is a way for the subject to escape from the present "demands of reality" into an imaginary past' (Laplanche and Pontalis, 1973: 112). Contrary to Jung's view – or at least to a Freudian interpretation of Jung's view – Freud referred specifically to 'whatever it has been impossible in the first instance to incorporate fully into a meaningful context. The traumatic event is the epitome of such unassimilated experience' (ibid.). So, whereas for Jung retroactive fantasizing (which takes place in the present moment) gives shape to the contents of the past in general, for Freud the concept of *Nachträglichkeit* suggests that in the present moment new meaning is attributed to the memory of a particular past traumatic event.

According to *Nachträglichkeit* the trauma is completed retrospectively, after the actual traumatic event takes place. In other words, meaning is given to a traumatic event at a later point in time. This is an important temporal mechanism in Freud's theory. Yet, even though he used this concept frequently, he did not analyze it extensively, and he did not connect it directly with time. It was Jacques Lacan who, while referring to the case of the Wolf Man (1953), resurfaced *Nachträglichkeit* and emphasized its significance for the understanding of the psyche's temporality.[3] Since Freud failed to provide a definition for the term, there are multiple meanings of *Nachträglichkeit* based on various temporal standpoints. This is reflected in ongoing debates regarding the translations of the term in English ('deferred action') as well as in French ('après-coup' = after the event/fact). There are contradictory interpretations regarding retranscription of the past event, which is either interpreted to imply 'causality, where the original event brings about subsequent pathological consequences' or 'the significance of the retrospective reinterpretation of repressed memory-traces' (Künstlicher 1994: 103–4). Also, 'These contradictory interpretations reflect Freud's own attitude and still colours the current debate about the translation issue' (ibid.). Freud's observed failure to provide a definition for *Nachträglichkeit* seems to be the reason for the dichotomy of opinions.

In an attempt to clarify the situation, Thomä and Cheshire suggested that Freud used the concept of *Nachträglichkeit* in two distinct ways that differ in their temporal direction:

> Sometimes [Freud] looks backwards from the point of view of the therapist reconstructing a phenomenal-developmental sequence of events and experiences, and sometimes he is looking forward as if through the eyes of the original traumatic event which is setting off a series of potentially pathogenic developments, and some of whose effects are going to be 'carried over' into the future.
>
> (Thomä and Cheshire 1991: 421)

Thus *Nachträglichkeit* implies two different directions in time:

a. A backward-looking/regressive movement, from the present moment *into the past* where the trauma originated: 'Something is deposited in the individual which is only activated later on ... Strachey's translation of Nachträglichkeit as "deferred action" ...conveys this particular meaning quite well' (Birksted-Breen 2003: 1512–13).
b. A forward-looking/progressive movement, from the past *into the present* moment, where the trauma is expressed: the event 'only takes on meaning retrospectively' (ibid.).

Authors who focus on the latter interpretation of the concept find the translation as 'deferred action' to be false: 'The essential difference between the verbs *nachtragen* and "to defer" consists in the fact that they express exactly opposite relations to time: the former is backward-looking (*nach* = "after", so after what?), whereas the latter is forward-looking (defer ... until when?)' (Perelberg 2006: 1203).

Modell indicated that '[*Nachträglichkeit*] does not mean "deferred action," as Strachey misleadingly translated it, but rather that subsequent experience results in a retranscription of memory or a retrospective attribution' (Modell 1990: 9). An event, with no particular significance at the time of its occurrence, acquires meaning subsequently, (i.e. 'after the fact') 'when there is a context within which [the event] can become meaningful' (Colman 2005: 650). This context, which can be provided in therapy or in a dream, supplies 'a symbolic or metaphorical shape for previously inchoate experience' (ibid.). The past event, according to this interpretation, is not traumatic in itself; the trauma requires two moments in time: the past event *and* the subsequent moment when meaning is made. According to this interpretation meaning-making was not postponed (which is what 'deferred action' implies); 'the reconstruction is not the discovery of an event which has been forgotten, but an attempt by means of reinterpretations to extract a new and hopefully more truthful understanding of the meaning of this event' (Künstlicher 1994: 116).

The difference in opinion lies in the significance attributed to the original event. In the case of 'deferred action' the event which caused the trauma is assumed to remain dormant until it is revisited in the future; its meaning exists but it is deferred. In the case of 'après coup' ('after the fact'), the past event is insignificant in its own right, until meaning is made at a later point in time. It is only after the initial event – literally, after the fact – that the original event becomes traumatic.

It has been indicated that a closer examination of these two possible interpretations suggests that they are not incompatible but are in fact complementary. For Sodré both directions of time are needed so that, 'The retranscription in the "past" of the internal world is experienced by the patient not as change, but as a capacity to see with greater clarity what was always there. A convincing new version of the past has to carry the belief that, 'in reality', this is how it has always been' (Sodré 2005: 8). In order for meaning to be made, one needs to examine the past event from the vantage point of the present, as well as reinterpret the present by taking the past event into account. Also, this approach suggests that it is not the event itself that is altered by retrospective examination; rather, 'Nachträglichkeit provides the memory, not the event, with traumatic significance and signifies a circular complementarity of both directions of time' (Eickhoff 2006: 1453). By retranscribing the past event at a later point in time it is integrated in one's life's narrative.

Birksted-Breen suggested that backward and forward movement in time are not merely two different aspects of temporality that are revealed through different uses of *Nachträglichkeit*, but rather they are interconnected:

> One movement cannot be separated from the other because *retroactive resignification is developmental progression.* For there to be progression there also has to be this kind of retrospective resignification. The forward movement necessitates a backward movement at the same time and, equally, the continual incorporation and restructuring of the past in the backward movement necessitates the ability to move forward.
>
> (Birksted-Breen 2003: 1509)

Backward movement made in order to reassign meaning is the reason why there is forward movement. Development happens because one goes backward; likewise, one cannot go backwards without being able to develop and acquire the means for meaning-making. Similarly, 'Multiple temporalities – progressive and regressive movements – take place together and condition each other reciprocally' (Perelberg 2006: 1204).

Notes

1 According to Strachey (the translator and editor of the *Standard Edition* of Freud's work), 'This seems to be the earliest explicit mention of the "timelessness" of the unconscious' (Freud 1901: 275). However, Freud made several remarks on the timelessness of the unconscious in earlier writings.

2 See Laplanche and Pontalis 1973; Schimek 1975; Thomä and Cheshire 1991; Künstlicher 1994; Faimberg 2005.

3 See Laplanche and Pontalis 1973; Thomä and Cheshire 1991; Künstlicher 1994; Barzilai 1997; Bernstein 2000; Birksted-Breen 2003; Faimberg 2005; Eickhoff 2006.

Bibliography

American Heritage Dictionary of the English Language (2003) Fourth Edition. Boston: Houghton Mifflin Company.

Arlow, J. A. (1986) 'Psychoanalysis and time', *Journal of the American Psychoanalytic Association,* 34: 507–28.

Augustine, Saint (1998) *Confessions*, trans. H. Chadwick. Oxford, New York: Oxford University Press.

Aziz, R. (1990) *C. G. Jung's Psychology of Religion and Synchronicity.* Albany: State University of New York Press.

Bambiniotis, G. (1998) *Λεξικό της Νέας Ελληνικής Γλώσσας* [Dictionary of Modern Greek Language]. Athens, Greece: Κέντρο Λεξικολογίας [Lexicology Centre] Ε.Π.Ε.

Barzilai, S. (1997) '"History is not the past": Lacan's critique of Ferenczi', *Psychoanalytic Review*, 84: 553–72.

Bergler, E. and Róheim, G. (1946) 'Psychology of time perception', *Psychoanalytic Quarterly*, 15: 190–206.

Bernstein, J. W. (2000) 'Making a memorial place: the photography of Shimon Attie', *Psychoanalytic Dialogues,* 10: 347–70.

Birksted-Breen, D. (2003) 'Time and the après-coup', *International Journal of Psycho-Analysis,* 84 (8): 1501–15.

Bishop, P. (2008) 'The timeliness and timelessness of the "archaic": analytical psychology, "primordial" thought, synchronicity', *Journal of Analytical Psychology*, 53: 501–23.

Bonaparte, M. (1940) 'Time and the unconscious', *International Journal of Psycho-Analysis*, 21: 427–68.

Borges, J. L. (1962) *Ficciones*. A. Kerrigan (ed.). New York: Grove Press.

—— (1974) 'Nueva refutacion del tiempo', in *Obras Completas de Jorge Luis Borges.* Buenos Aires: Emece Editores.

Brenner, E. M. (1990) 'Gnosticism and psychology: Jung's *Septem Sermones ad Mortuos*', *Journal of Analytical Psychology*, 35 (4): 397–419.

Bright, G. (1997) 'Synchronicity as a basis of analytic attitude', *Journal of Analytical Psychology*, 42: 613–35.

Brizzi, S. (1998) 'The metaphor of time in psychoanalytic technique', *Journal of the American Academy of Psychoanalysis*, 26: 51–68.

Brontë, E. (1953) *Wuthering Heights*. London and Glasgow: Collins.

Calhoun, C. J. (1976) 'Continuity and change: the significance of time in the organization of experience', *International Journal of Psychoanalysis*, 3: 291–304.

Cambray, J. (2009) *Synchronicity: Nature and Psyche in an Interconnected Universe*. College Station, TX: Texas A & M University Press.

Capek, M. (ed.) (1976) *The Concepts of Space and Time; Their Structure and Their Development*. Boston Studies in the Philosophy of Science, Volume XXII, Holland and USA: D. Reidel Publishing Company.

Card, C. R. (1991a) 'The archetypal view of C. G. Jung and Wolfgang Pauli, Part I', *Psychological Perspectives*, 24: 19–33.

—— (1991b) 'The archetypal view of C. G. Jung and Wolfgang Pauli, Part II: the archetypal hypothesis in psychology and physics', *Psychological Perspectives*, 25: 52–69.

Carroll, L. (1960) *Alice's Adventures in Wonderland and Through the Looking-Glass*. New York: Signet Classic.

Carter, L. and West, M. (2008) Book review of *The Innermost Kernel: Depth Psychology and Quantum Mechanics; Wolfgang Pauli's dialogue with C. G. Jung*, *Journal of Analytical Psychology*, 53: 127–140.

Cohn, F. S. (1957) 'Time and the ego', *Psychoanalytic Quarterly*, 26: 168–89.

Colman, W. (2005) 'Sexual metaphor and the language of unconscious phantasy', *Journal of Analytical Psychology*, 50: 641–60.

Crastnopol, M. (2002) 'The dwelling places of self-experience', *Psychoanalytic Dialogues*, 12: 259–84.

Cummings, E. E. (1997) *E. E. Cummings: Selected Poems 1923–1958*. London: Faber and Faber.

Davies, P. (2002, September) 'The Mysterious Flow', *Scientific American*, 287: 32–7.

De Rola, S. K. (1997) *Alchemy, the Secret Art*. London: Thames and Hudson.

De Simone, G. (1997) *Ending Analysis: Theory and Technique*. London: Karnac.

Dehing, J. (1990) 'Jung and knowledge: from gnosis to praxis', *Journal of Analytical Psychology*, 35: 377–96.

Eickhoff, F. (2006) 'On Nachträglichkeit: the modernity of an old concept', *International Journal of Psycho-Analysis*, 87: 1453–69.

Eliot, T. S. (1944) *Four Quartets*. London: Faber and Faber.

Ellis, M. L. (2008) *Time in Practice: Analytical Perspectives on the Times of Our Lives*. London: Karnac.

Faimberg, H. (2005) 'Après-coup', *International Journal of Psychoanalysis*, 86: 1–6.

Frank, G. (1996) 'Beliefs and their vicissitudes', *Psychoanalytic Psychology*, 13: 421–31.

Freud, S. (1894) *Neuro-psychoses of Defence*, in J. Strachey (ed. and trans.), *The Standard Edition of the Complete Psychological Works of Sigmund Freud* [hereafter S.E.], Vol. 3 (1893–99). London: Vintage, The Hogarth Press and the Institute of Psychoanalysis.

—— (1896) *The Aetiology of Hysteria*, S.E., Vol. 3 (1893–99).

—— (1900) *The Interpretation of Dreams*. New York: Basic Books Inc.
—— (1901) *Psychopathology in Everyday Life: Forgetting, Slips of the Tongue, Bungled Actions, Superstitions and Errors*, S.E., Vol. 6.
—— (1914) *On Narcissism: An Introduction*, S.E., Vol. 14 (1914–16).
—— (1915) *The Unconscious*, S.E., Vol. 14 (1914–16).
—— (1918) *From the History of an Infantile Neurosis (Wolf-man Case Study)*, S.E., Vol. 17 (1917–19).
—— (1920) *Beyond the Pleasure Principle*, S.E., Vol. 18 (1920–22).
—— (1923) *The Ego and the Id*, S.E., Vol. 19 (1923–25).
—— (1926) *Inhibitions, Symptoms and Anxiety*, S.E., Vol. 20 (1925–26).
—— (1930) *Civilization and its Discontents*. S.E., Vol. 21 (1927–31).
—— (1933) *New Introductory Lectures on Psycho-Analysis*, S.E., Vol. 22 (1932–36).
—— (1950) *Project for a Scientific Psychology*, S.E., Vol. 1 (1886–99).
—— (1954) *The Origins of Psychoanalysis; Letters to Wilhelm Fliess; Drafts and Notes (1887–1902)*. New York: Basic Books.
Gammon, M. (1973) '"Window into eternity": archetype and relativity', *Journal of Analytical Psychology*, 18 (1): 11–14.
Gell, A. (2001) *The Anthropology of Time: Cultural Constructions of Temporal Maps and Images*. Oxford and Washington, DC: Berg.
Gieser, S. (2005) *The Innermost Kernel; Depth Psychology and Quantum Physics: Wolfgang Pauli's Dialogue with C. G. Jung*. Berlin, Heidelberg, New York: Springer.
Gifford, S. (1960) 'Sleep, time and the early ego', *Journal of American Psychoanalytic Association*, 8: 5–42.
—— (1980) '"The prisoner of time": some developmental aspects of time perception in infancy, sensory isolation, and old age', *Annual Psychoanalysis*, 8: 131–54.
Gordon, R. (1977) 'Death and creativity: a Jungian approach', *Journal of Analytical Psychology*, 22 (2): 106–24.
Green, A. (2002) *Time in Psychoanalysis*, trans. A. Weller. London and New York: Free Association Books.
—— (2009) 'From the ignorance of time to the murder of time. From the murder of time to the misrecognition of temporality in psychoanalysis', in L. G. Fiorini and J. Canestri (eds), *The Experience of Time in Psychoanalysis: Psychoanalytic Perspectives*. London: Karnac.
Grivet-Shillito, M. L. (1999) 'Carl Gustav before he became Jung', *Journal of Analytical Psychology*, 44: 87–100.
Harrison, R. P. (2008) *Gardens: An Essay on the Human Condition*. Chicago and London: University of Chicago Press.
Hartocollis, P. (1974) 'Origins of time; a reconstruction of the ontogenetic development of the sense of time based on object-relations theory', *Psychoanalytic Quarterly*, 43: 243–61.
Hillman, J. (1975) 'Abandoning the child' (1971), in *Loose Ends: Primary Papers in Archetypal Psychology*. Dallas, TX: Spring Publications.

Hoeller, S. A. (1982) *The Gnostic Jung and the Seven Sermones to the Dead*. Wheaton, IL: Theosophical.

Horne, M, Sowa, A. and Isenman, D. (2000) 'Philosophical assumptions in Freud, Jung and Bion: questions of causality', *Journal of Analytical Psychology*, 45: 109–21.

Hubback, J. (1966) 'VII Sermones ad Mortuos', *Journal of Analytical Psychology*, 11 (2): 95–112.

Jacobson, E. (1954) 'Contribution to the metapsychology of psychotic identification', *Journal of the American Psychoanalytic Association*, 2: 239–62.

Jung, C. G. (1897) *The Zofingia Lectures 83–142: Some Thoughts on Psychology*, Sir H. Read, M. Fordam and G. Adler (eds), trans. J. V. Heurck. London: Routledge & Kegan Paul.

—— (1911–12/1952) *Symbols of Transformation*, in Sir H. Read, M. Fordam and G. Adler (eds), W. McGuire (executive ed.), trans. R. F. C. Hull, *The Collected Works of C. G. Jung* [hereafter C.W.], Vol. 5. London: Routledge & Kegan Paul.

—— (1913) *Freud and Psychoanalysis: The Theory of Psychoanalysis*, in C.W., Vol. 4.

—— (1914) *The Psychogenesis of Mental Disease: On Psychological Understanding*, in C.W., Vol. 3.

—— (1916) *Two Essays on Analytical Psychology; Appendices: the Structure of the Unconscious*, in C.W., Vol. 7.

—— (1916/1917) *Preface to 'Collected Papers on Analytical Psychology'*, in C.W., Vol. 4.

—— (1917/1926/1943) *Two Essays on Analytical Psychology: On the Psychology of the Unconscious*, in C.W., vol. 7.

—— (1921) *Psychological Types*, C.W., vol. 6.

—— (1928a) *The Structure and Dynamic of the Psyche: On Psychic Energy*, in C.W., Vol. 8.

—— (1928b) *Two Essays on Analytical Psychology: The Relations Between the Ego and the Unconscious*, in C.W., Vol. 7.

—— (1928/1931) *The Structure and Dynamics of the Psyche: Analytical Psychology and Weltanschauung*, in C.W., Vol. 8.

—— (1929) *The Significance of Constitution and Heredity in Psychology*, in C.W., Vol. 8.

—— (1930) *The Practice of Psychotherapy: Some Aspects of Modern Psychotherapy*, in C.W., Vol. 16.

—— (1931) *Archaic Man*, in C.W., Vol. 10.

—— (1933/1934) *Civilization in Transition; The Meaning of Psychology for Modern Man*, in C.W., Vol. 10.

—— (1934) *The Soul and Death*, in C.W., Vol. 8.

—— (1934/1954) *Archetypes of the Collective Unconscious*, in C.W., Vol. 9i.

—— (1935) *The Tavistock Lectures*, in C.W., Vol. 18.

—— (1936) *Psychology and Alchemy: Individual Dream Symbolism in Relation to Alchemy*, in C.W., Vol. 12.

—— (1937) *Psychology and Alchemy: Religious Ideas in Alchemy*, in C.W., Vol. 12.

—— (1938/1940) *Psychology and Religion: The Terry Lectures*, in C.W., Vol. 11.

—— (1938/1954) *Alchemical Studies: The Vision of Zosimos*, in C.W., Vol. 13.
—— (1939) *The Symbolic Life*, in C.W., Vol. 18.
—— (1939/1954) *Psychology and Religion: West and East; Psychological Commentary on 'The Tibetan Book of the Great Liberation'*, in C.W., Vol. 11.
—— (1940) *Archetypes and the Collective Unconscious: Psychology of the Child Archetype*, in C.W., Vol. 9i.
—— (1942/1948). *Psychology and Religion: West; A psychological Approach to the Dogma of the Trinity*, in C.W., Vol. 11.
—— (1942/1954) *Psychology and Religion: West; Transformation Symbolism in the Mass*, in C.W., Vol. 11.
—— (1947/1954) *The Structure and Dynamics of the Psyche: On the Nature of the Psyche*, in C.W., Vol. 8.
—— (1948) *On Occultism: Psychology and Spiritualism*, in C.W., Vol. 18.
—— (1949) *Forward to Neumann: Depth Psychology and a New Ethic*, in C.W., Vol. 18.
—— (1950) *The Archetypes and the Collective Unconscious: Concerning Mandala Symbolism*, in C.W., Vol. 9i.
—— (1951a) *The Practice of Psychotherapy: Fundamental Questions of Psychotherapy*, in C.W., Vol. 16.
—— (1951b) *Aion*, in C.W., Vol. 9ii.
—— (1952a) *Answer to Job*, in C.W., Vol. 11.
—— (1952b). *Religion and Psychology: A Reply to Martin Buber*, in C.W., Vol. 18.
—— (1952c) *The Structure and Dynamic of the Psyche; Synchronicity: An Acausal Connecting Principle*, in C.W., Vol. 8.
—— (1954) *Psychology and Religion: On Resurrection*, in C.W., Vol. 18.
—— (1955) *Letters on Synchronicity: to Michael Fordham*, in C.W., Vol. 18.
—— (1955–56) *Mysterium Coniuctionis*, C.W., Vol. 14.
—— (1957) *The Undiscovered Self: Present and Future*, in C.W., Vol. 10.
—— (1958) *Civilization in Transition: A Psychological View of Conscience*, in C.W., Vol. 10.
—— (1961) *Memories, Dreams, Reflections*. Toronto, Canada: Vintage Books.
—— (1973) *Letters I: 1906–1950*, G. Adler (ed.), Bollingen Series XCV1. Princeton, NJ: Princeton University Press.
—— (1976) *Letters II: 1951–1961*, G. Adler (ed.), Bollingen Series XCV2. Princeton, NJ: Princeton University Press.
—— (1984) *Dream Analysis: Notes of the Seminars Given in 1928–1930 by C. G. Jung*, W. McGuire (ed.), Bollingen Series XCIX. Princeton, NJ: Princeton University Press.
—— (2008) *Children's Dreams: Notes from the Seminar Given in 1936–1940*, L. Jung and M. Meyer-Grass (eds), Philemon Series. Princeton, NJ: Princeton University Press.
—— (2009) *The Red Book: Liber Novus*, S. Shamdasani (ed.). New York, London: Norton.

Kafka, J. S. (1972) 'The experience of time', *Journal of American Psychoanalytic Association*, 20: 650–67.

Kerényi, C. (2006) *The Gods of the Greeks*. London: Thames and Hudson.

King, P. (2005) 'Time and sense of identity', in *Time Present and Time Past: Selected Papers of Pearl King*. London: Karnac.

Klein, S. (2006) *Time: A User's Guide; Making Sense of Life's Scarcest Commodity*. London: Penguin Books.

Künstlicher, R. (1994) '"Nachträglichkeit": the intermediary of an unassimilated impression and experience', *Scandinavian Psychoanalytic Review*, 17: 101–18.

Laplanche, J. and Pontalis, J. B. (1973) *The Language of Psycho-Analysis*. London: Karnac.

Lightman, A. (1993) *Einstein's Dreams: A Novel*. New York: Warner Books.

Lindorff, D. (1995) 'Psyche, matter and synchronicity; a collaboration between C. G. Jung and Wolfgang Pauli', *Journal of Analytical Psychology*, 40: 571–86.

Loewald, H. W. (1962) 'The superego and the ego-ideal II: superego and time', *International Journal of Psychoanalysis*, 43: 264–68.

Main, R. (2004) *The Rupture of Time: Synchronicity and Jung's Critique of Modern Western Culture*. Hove and New York: Brunner-Routledge, Taylor & Francis Group.

—— (2007) *Revelations of Change: Synchronicity as Spiritual Experience*. Albany: State University of New York Press.

Masler, E. G. (1973) 'The subjective perception of two aspects of time: duration and timelessness', *International Journal of Psychoanalysis*, 54: 425–9.

Mathers, D. (2001) *An Introduction to Meaning and Purpose in Analytical Psychology*. East Sussex, PH: Brunner-Routledge.

Meier, C. A. (ed.) (2001) *Atom and Archetype; the Pauli/Jung Letters, 1932–1958*. London: Routledge; New Jersey: Princeton University Press.

Meissner, W. (2007) *Time, Self and Psychoanalysis*. Lanham, Boulder, New York, Toronto, Plymouth: Jason Aronson.

Michaels, A. (1997) *Fugitive Pieces*. New York: Vintage.

Miller, A. (2009) *Deciphering the Cosmic Number: The Strange Friendship of Wolfgang Pauli and Carl Jung*. New York, London: W. W. Norton & Company.

Milton, J. (1905) *Paradise Lost*. London: Routledge and Sons.

Mitchell, S. A. (1993) *Hope and Dread in Psychoanalysis*. New York: Basic Books.

Modell, A. H. (1990) *Other Times, Other Realities. Towards a Theory of Psychoanalytic Treatment*. Cambridge, MA: Harvard University Press.

Morgan, H. (2000) 'The new physics through a Jungian perspective', in E. Christopher and H. Solomon (eds), *Jungian Thought in the Modern World*. London: Free Association Books.

Morris, J. (1983) 'Time experience and transference', *Journal of American Psychoanalytic Association*, 31: 651–76.

Namnum, A. (1972) 'Time in psychoanalytic technique', *Journal of the American Psychoanalytic Association*, 20: 736–50.

Neumann, C. (1976) 'On the necessity of the absolute frame of reference', in M. Capek (ed.), *The Concepts of Space and Time: Their Structure and Their Development*, Boston Studies in the Philosophy of Science, Vol. XXII. Holland and USA: D. Reidel Publishing Company.

Newton, I. (1952) *Newton's Philosophy of Nature: Selections from his Writings*, S. H. Thayer (ed.). New York: S. H. Hafner Publishing Company.

Nichols, S. (1980) *Jung and the Tarot: An Archetypal Journey*. San Francisco, Newburyport: Weiser Books.

Noel-Smith, K. (2002) 'Time and space as "necessary forms of thought"', *Free Associations*, 9 (C): 394–442.

Nunberg, H. and Federn, E. (eds) (1974) *Minutes of the Vienna Psychoanalytic Society, Vol. III, 1910–11*. New York: International Universities Press.

Papadopoulos, R. K. (1980) 'The dialectic of the other in the psychology of C. G. Jung: a metatheoretical investigation', unpublished thesis, University of Cape Town, South Africa.

—— (2006) 'Jung's epistemology and methodology' in R. K. Papadopoulos (ed.), *The Handbook of Jungian Psychology: Theory, Practice and Applications*. London and New York: Routledge.

Peat, D. (1988) 'Divine contenders: Wolfgang Pauli and the symmetry of the world', *Psychological Perspectives*, 19: 14–23.

Perelberg, R. J. (2006) 'The controversial discussions and après-coup', *International Journal of Psychoanalysis,* 87: 1199–220.

—— (2008) *Time, Space and Phantasy*. East Sussex, New York: Routledge.

Plessner, H. (1957) 'The relation of time to death' in J. Campbell (ed.), *Man and Time: Papers from the Eranos Yearbook*, Bollingen Series, 30, Vol. 3. New York: Pantheon Books Inc.

Pollock, G. H. (1971) 'On time, death, and immortality', *Psychoanalytic Quarterly*, 40: 435–46.

Priel, B. (1997) 'Time and self: on the intersubjective construction of time', *Psychoanalytic Dialogues*, 7: 431–50.

Progoff, I. (1973) *Jung, Synchronicity, and Human Destiny; C. G. Jung's Theory of Meaningful Coincidence*. New York: Julian Press, Crown Publishers.

Puech, H. C. (1957) 'Gnosis and time', in J. Campbell (ed.), *Man and Time: Papers from the Eranos Yearbooks*, Bollingen Series, 30, Vol. 3. New York: Pantheon Books Inc.

Reyner, E. and Tuckett, D. (eds) (2005) 'An introduction to Matte-Blanco's reformulation of the Freudian unconscious and his conceptualization of the internal world', in I. Matte-Blanco (ed.), *Thinking, Feeling, and Being: Clinical Reflections on the Fundamental Antinomy of Human Beings and World*. London and New York: Routledge.

Rilke, R. M. (1960) *Sonnets to Orpheus*, trans. C. F. MacIntyre. Berkeley, Los Angeles, London: University of California Press.

Rose, J. (1997) 'Distortions of time in the transference: some clinical and theoretical implications', *International Journal of Psychoanalysis*, 78: 453–68.

Rudolph, K. (1987) *Gnosis: The Nature and History of Gnosticism*. San Francisco: Harper.

Russell, G. (1931) *A. E., Voices of the Stones*. New York: Macmillan.

Sabbadini, A. (1989) 'Boundaries of timelessness: some thoughts about the temporal dimension of the psychoanalytic space', *International Journal of Psychoanalysis*, 70: 305–13.

Samuels, A. (1985) *Jung and the Post-Jungians*. London, New York: Routledge.

Samuels, A., Shorter, B. and Plaut, F. (2003) *A Critical Dictionary of Jungian Analysis*. East Sussex, New York: Brunner-Routledge.

Schafer, R. (1972) 'Internalization: process or fantasy?', *Psychoanalytic Study of the Child*, 27: 411–36.

Schimek, J. G. (1975) 'The interpretation of the past: childhood trauma, psychical reality, and historical truth', *Journal of the American Psychoanalytic Association*, 23: 845–65.

Schneemelcher, W. (ed.) (1991) *New Testament Apocrypha, Vol 1: Gospels and Related Writings*, R. Wilson (trans.). Louisville, KT: Westminster John Knox Press.

Segal, R. A. (2007) 'Jung and Lévy-Bruhl', *Journal of Analytical Psychology*, 52: 635–58.

Siegelman, E. Y. (1990) 'Metaphors of the therapeutic encounter', *Journal of Analytical Psychology*, 35: 175–91.

Slochower, J. (1999) 'Interior experience within analytic process', *Psychoanalytic Dialogues*, 9: 789–809.

Sodré, I. (2005) '"As I was walking down the stair, I saw a concept which wasn't there…": or, après-coup: a missing concept?', *International Journal of Psychoanalysis*, 86: 7–10.

Spengler, O. (1991) *The Decline of the West*, trans. C. F. Atkinson. New York and Oxford: Oxford University Press.

Stein, M. (2003) *Jung's Map of the Soul*. Chicago, La Salle, IL: Open Court.

Terr, L. C. (1984) 'Time and trauma', *Psychoanalytic Study of the Child*, 39: 633–65.

Thomä, H. and Cheshire, N. (1991) 'Freud's Nachträglichkeit and Strachey's "deferred action": trauma, constructions and the direction of causality', *International Review of Psychoanalysis*, 18: 407–27.

Von Franz, M. L. (1974) *Number and Time: Reflections Leading towards a Unification of Depth Psychology and Physics*. Evanston, IL: Northwestern University Press.

—— (1980) *Alchemy: An Introduction to the Symbolism and the Psychology*. Toronto: Inner City Books.

—— (1983) 'Introduction to the Zofingia Lectures', in Sir H. Read, M. Fordam and G. Adler (eds), trans. J. V. Heurck, *The Zofingia Lectures, C. G. Jung*. London: Routledge & Kegan Paul.

—— (1992) *Psyche and Matter*. Boston and London: Shambala.

Williams, M. (1958) 'The fear of death; part 1: the avoidance of the fear of death', *Journal of Analytical Psychology*, 3 (2): 157–65.

Yiassemides, A. (2011) 'Chronos in synchronicity: manifestations of the psychoid reality', *Journal of Analytical Psychology*, 56: 451–70.

Zabriskie, B. (1995) 'Jung and Pauli: a subtle asymmetry', *Journal of Analytical Psychology*, 40: 531–53.

—— (2001) 'Jung and Pauli: a meeting of rare minds', in C. A. Meier (ed.), *Atom and Archetype: The Pauli/ Jung Letters, 1932–1958*. London: Routledge; New Jersey: Princeton University Press.

Zavala, J. F. (1982) 'Synchronicity and the Mexican divinatory calendar *Tonalamatl*', *Quadrant* 15 (1): 55–70.

Index

9780415810432